# Applications of Fluoropolymer Films

PLASTICS DESIGN LIBRARY (PDL) PDL HANDBOOK SERIES

**Series Editor**: Sina Ebnesajjad, PhD (sina@FluoroConsultants.com) President, FluoroConsultants Group, LLC Chadds Ford, PA, USA
http://www.FluoroConsultants.com

The **PDL Handbook Series** is aimed at a wide range of engineers and other professionals working in the plastics industry, and related sectors using plastics and adhesives.

PDL is a series of data books, reference works and practical guides covering plastics engineering, applications, processing, and manufacturing, and applied aspects of polymer science, elastomers and adhesives.

**Recent titles in the series**

Recycling of Flexible Plastic Packaging 1, Niaounakis, Michael (ISBN: 9780128163351)
Plasticizers Derived from Post-Consumer PET 1, Langer, Ewa (ISBN: 9780323462006)
Polylactic Acid 2, Sin, Lee Tin (ISBN: 9780128144725)
Durability and Reliability of Polymers and Other Materials in Photovoltaic Modules 1, Yang, Hsinjin; French, Roger; Bruckman, Laura (ISBN: 9780128115459)
Fluoropolymer Additives 2, Ebnesajjad, Sina; Morgan, Richard (ISBN: 9780128137840)
The Effect of UV Light and Weather on Plastics and Elastomers 4, McKeen, Larry (ISBN: 9780128164570)
PEEK Biomaterials Handbook 2, Kurtz, Steven (ISBN: 9780128125243)
Hydraulic Rubber Dam, Thomas et al. (ISBN: 9780128122105)
Electrical Conductivity in Polymer-based Composites, Taherian & Kausar (ISBN: 9780128125410)
Plastics to Energy, Al-Salem (ISBN: 9780128131404)
Recycling of Polyethylene Terephthalate Bottles, Thomas et al. (ISBN: 9780128113615)
Dielectric Polymer Materials for High-Density Energy Storage, Dang (ISBN: 9780128132159)
Thermoplastics and Thermoplastic Composites, Biron (ISBN: 9780081025017)
Recycling of Polyurethane Foams, Thomas et al. (ISBN: 9780323511339)
Introduction to Plastics Engineering, Shrivastava (ISBN: 9780323395007)
Chemical Resistance of Thermosets, Baur, Ruhrberg & Woishnis (ISBN: 9780128144800)
Phthalonitrile Resins and Composites, Derradji, Jun & Wenbin (ISBN: 9780128129661)
The Effect of Sterilization Methods on Plastics and Elastomers, 4e, McKeen (ISBN: 9780128145111)
Polymeric Foams Structure-Property-Performance, Obi (ISBN: 9781455777556)
Technology and Applications of Polymers Derived from Biomass, Ashter (ISBN: 9780323511155)
Fluoropolymer Applications in the Chemical Processing Industries, 2e, Ebnesajjad & Khaladkar (ISBN: 9780323447164)
Reactive Polymers, 3e, Fink (ISBN: 9780128145098)
Service Life Prediction of Polymers and Plastics Exposed to Outdoor Weathering, White, White & Pickett, (ISBN:9780323497763)
Polylactide Foams, Nofar & Park (ISBN: 9780128139912)
Designing Successful Products with Plastics, Maclean-Blevins (ISBN: 9780323445016)
Waste Management of Marine Plastics Debris, Niaounakis, (ISBN: 9780323443548)
Film Properties of Plastics and Elastomers, 4e, McKeen, (ISBN: 9780128132920)
Anticorrosive Rubber Lining, Chandrasekaran (ISBN: 9780323443715)
Shape-Memory Polymer Device Design Safranski & Griffis, (ISBN: 9780323377973)
A Guide to the Manufacture, Performance, and Potential of Plastics in Agriculture, Orzolek, (ISBN: 9780081021705)
Plastics in Medical Devices for Cardiovascular Applications, Padsalgikar, (ISBN: 9780323358859)
Industrial Applications of Renewable Plastics, Biron (ISBN: 9780323480659)
Permeability Properties of Plastics and Elastomers, 4e, McKeen, (ISBN: 9780323508599)
Expanded PTFE Applications Handbook, Ebnesajjad (ISBN: 9781437778557)
Applied Plastics Engineering Handbook, 2e, Kutz (ISBN: 9780323390408)
Modification of Polymer Properties, Jasso-Gastinel & Kenny (ISBN: 9780323443531)
The Science and Technology of Flexible Packaging, Morris (ISBN: 9780323242738)
Stretch Blow Molding, 3e, Brandau (ISBN: 9780323461771)
Chemical Resistance of Engineering Thermoplastics, Baur, Ruhrberg & Woishnis (ISBN: 9780323473576)
Chemical Resistance of Commodity Thermoplastics, Baur, Ruhrberg & Woishnis (ISBN: 9780323473583)

To submit a new book proposal for the series, or place an order, please contact Edward Payne, Acquisitions Editor at edward.payne@elsevier.com

# Applications of Fluoropolymer Films

## Properties, Processing, and Products

**Jiri George Drobny**

Drobny Polymer Associates

ELSEVIER

William Andrew
Applied Science Publishers

William Andrew is an imprint of Elsevier
The Boulevard, Langford Lane, Kidlington, Oxford, OX5 1GB, United Kingdom
50 Hampshire Street, 5th Floor, Cambridge, MA 02139, United States



**British Library Cataloguing-in-Publication Data**
A catalogue record for this book is available from the British Library

**Library of Congress Cataloging-in-Publication Data**
A catalog record for this book is available from the Library of Congress

ISBN: 978-0-12-816128-9

For Information on all William Andrew publications
visit our website at https://www.elsevier.com/books-and-journals

*Publisher:* Matthew Deans
*Acquisition Editor:* Edward Payne
*Editorial Project Manager:* Peter Adamson
*Production Project Manager:* Kamesh Ramajogi
*Cover Designer:* Miles Hitchen

Typeset by MPS Limited, Chennai, India

# Dedication

Dedicated to my family for their love and support

# Contents

# Preface

Fluoropolymer films are high-performance films based on fluoroplastic polymers, such as polytetrafluoroethylene, and others. These films as a class exhibit low coefficient of friction, chemical inertness, exceptional dielectric properties, weather and UV resistance, excellent optical properties, negligible moisture absorption, and outstanding performance at very high temperatures.

Fluoropolymer films are used in many markets, although none of them is used in large quantities. Current market demand for specialty fluoropolymer films is approximately $150 million and expected to rise 5.4% per year to $177 million in 2023. The growth is expected to be among the fastest of any major specialty resin, driven largely by expanding use of fluoropolymers mainly in photovoltaic modules and also in other markets, including fuel cells, health care, and specialty packaging materials.

Recently published books on the general subject of films are *Film Properties of Plastics and Elastomers, Fourth Edition*, by L.W. McKeen (Elsevier 2017); *Manufacture and Novel Applications of Multilayer Polymer Films*, by D. Langhe and M. Ponting (Elsevier 2016); and *Science and Technology of Flexible Packaging: Multilayer Films from Resins and Process to End Use* by B.A. Morris (Elsevier, 2017). There is only one book on the subject of fluoropolymer films, namely the very comprehensive and well-written monograph *Polyvinyl Fluoride: Technology and Applications of PVF*, by S. Ebnesajjad, (Elsevier, 2015). Clearly, considering the importance and the large number of types of fluoropolymer films currently being used, the decision was made by the Elsevier publishers to publish a book that would be dedicated to that subject.

This book is written with the intention to be useful for industrial and business practitioners and have its main focus on the practical use. We hope that it will provide relevant information about the essential technology of materials, processing and properties and applications of every type of fluoropolymer films. The main focus is on applications. In order to accomplish that, the publication is divided into the following three parts:

Part I: Materials, Technology, and Properties
Part II: Typical Properties of and Applications for Fluoropolymer Films
Part III: Commercial Grades of Fluoropolymer Films and Their Applications

The Part I is written so that it does not require any deep understanding of polymer chemistry, since we limited the discussion to industrial processes for the production of individual polymers and then to their processing into films. We provided ample references, and in addition to that we listed all known published books covering the subjects of fluoropolymers, polymeric films, and film processing technology, and testing and applications as the bibliography. A comprehensive glossary of terms is also included. An entire chapter is dedicated to testing and current standard for the testing of thermoplastic films in general and fluoroplastic films in particular.

The Part II covers typical properties and examples of applications for every single type of fluoropolymer films. Each type of film is presented as a separate chapter in order to focus on all important details.

The Part III is a comprehensive description of essentially all commercial grades of fluoropolymer films. We did a thorough research of available manufacturers, suppliers, and grades. Eventually, we decided to list the major manufacturers and suppliers and essential grades. The sources are data published by individual companies. It should be noted that not all companies producing and supplying fluoropolymer films use data in a uniform fashion; for example, often different units are shown. Where possible, units have been changed for the sake of uniformity. In addition, sometimes the number of different grades are very large so a few typical ones have been selected. For additional information the reader is advised to study the entire range of products listed on the website of a given company. Our intention is to provide as much useful information as possible for the reader, but it is in no way to guarantee accuracy and to claim that the data are specific. The reader is advised to contact the manufacturer, supplier, and/or distributor for that.

Many thanks are due to my friend and colleague Dr. Sina Ebnesajjad, President of FluoroConsultants Group for encouragement, help, and valuable advice and to Edward Payne and Dr. Peter Adamson from Elsevier for outstanding support and help during the preparation and writing of the manuscript and to Corinne Gangloff from Freedonia

Group for supplying market data and forecasts. Additional credit is due to the Elsevier production team managed by Kamesh Ramajogi for bringing this work to fruition.

**Jiri George Drobny**

*Merrimack, NH, Boulder, CO, and Prague, Czech Republic*

# PART I

# Materials, Technology, and Properties

# 1 Introduction

Fluoropolymer films are high-performance films based on fluoroplastic polymers, such as polytetrafluoroethylene (PTFE), known under its trade name TEFLON and others. These films as a class exhibit low coefficient of friction, chemical inertness, exceptional dielectric properties, weather/UV resistance, excellent optical properties, negligible moisture absorption, and outstanding performance at temperature extremes. Some of them can be thermoformed, laminated, heat-sealed, die-stamped, and oriented for use in a wide variety of applications. The key application industries for these films include packaging, construction, automotive, electronics, electrical, health care, and aerospace and typical applications include anticorrosive linings, composite part mold release, industrial roll covers, circuitry, pharmaceutical cap liners, sterile packaging, cable insulation, hot-melt adhesive, microphone electret membranes, photovoltaic cell glazing (backsheet), antigraffiti coverings, erasable surface coverings, automotive interiors, fuel hose permeation barrier, hot-melt adhesive, and more.

Fluoropolymer films are used in a wide array of markets, although none of them is used in large quantities. Demand for specialty fluoropolymer films is expected to rise 5.4% per year to $177 million in 2023 [1]. The growth will be among the *fastest of any major specialty resin*, driven largely by expanding the use of fluoropolymers in *photovoltaic modules*. The demand will be also driven by other markets, including fuel cells, health care, and packaging materials. However, the gains will be limited by less expensive alternative materials than the relatively high-priced fluoropolymer materials. Thus fluoropolymer films will be used only where their high barrier properties, chemical resistance, or toughness are essential [1]. As it stands now, the share of fluoropolymer films amounts to 1.8%–2% of the total specialty film demand.

Polyvinyl fluoride (PVF) is currently the leading fluoropolymer used in film applications. PVF and PVDF are widely used in backsheets of photovoltaic modules. In fact majority of backsheets produced in 2018 contained at least one layer of fluoropolymer-based film [1]. There is a possibility that the share of PVF-based backsheets may decline despite increasing demand for solar energy products, as newer technologies gain market share [1].

**Applications of Fluoropolymer Films.** DOI: https://doi.org/10.1016/B978-0-12-816128-9.00001-5

US demand for all specialty films is forecast to rise 4.0% per year to $9.0 billion in 2023. The gains will be fueled by many factors, including demand for high-performance films with barrier properties, growth construction activity, acceleration in packaging production, and food manufacturing. Key markets for barrier films will be food packaging, pharmaceutical blister packaging applications, particularly fluoropolymer, and nylon films in multilayer constructions [1]. Details summarizing the relevant data are in Tables 1.1 and 1.2.

**Table 1.1 Fluoropolymer Film Demand by Function, 2008–23 (Millions US Dollars) [1].**

| | Year | | | |
|---|---|---|---|---|
| **Item** | **2008** | **2013** | **2018** | **2023** |
| Fluoropolymer demand | 96 | 111 | 136 | 177 |
| Barrier films | 31 | 47 | 72 | 97 |
| Light control films | 20 | 17 | 14 | 19 |
| Conductive and insulative films | 15 | 9 | 6 | 7 |
| Other functions | 30 | 38 | 44 | 54 |
| Total all specialty film demand | 5296 | 6199 | 7420 | 9010 |
| Fluoropolymer films (%) | 1.8 | 1.8 | 1.8 | 2.0 |

**Table 1.2 Specialty Fluoropolymer Films Demand by Function, 2008–13 (Compound Annual Growth) [1].**

| | Compound Annual Growth (%) | | |
|---|---|---|---|
| **Item** | **2008–13** | **2013–18** | **2018–23** |
| Specialty fluoropolymer demand | 2.9 | 4.1 | 5.4 |
| Barrier films | 8.7 | 8.9 | 6.1 |
| Light control films | − 3.2 | − 3.8 | 6.3 |
| Conductive and insulative films | − 9.7 | − 7.8 | 3.1 |
| Other functions | 4.8 | 3.0 | 4.2 |
| Total all specialty film demand | 3.2 | 3.7 | 4.0 |
| Fluoropolymer films (%) | 0.0 | 0.0 | 2.1 |

## 1.1 Definitions

### 1.1.1 Polymeric Films and Sheets

Polymeric films are defined as thin continuous materials typically up to 200 μm (0.008 in.) thick. Plastic materials thicker than that are referred to as sheets. Polymeric films can be manufactured from various resins, each with its own unique physical properties that are differently suited to different applications. Besides all the different materials it can be made from, a polymeric film can be clear, colored, smooth, rough, functionally embossed, opaque, or semitransparent.

Polymeric films are essentially made from thermoplastic resins using the following manufacturing methods:

- *Film extrusion* using a flat die and subsequent cooling and wounding up on a roll.
- *Cast film extrusion* is an extrusion method in which the polymer melt is cooled or quenched and then wound up on a roll.
- *Blown film extrusion* using a special extrusion die to extrude a tube. The melted tube is then inflated to the required circumference and subsequently cooled down and collapsed between two steel rolls as two sheets.
- *Coextrusion* produces films composed from two or more different polymeric layers.
- *Calandering*, in which a polymeric melt is shaped between two steel rolls.
- *Skiving* is peeling of the film or sheet from a solid roll (billet) in a similar fashion as a wood veneer.
- *Film casting* is a deposition of a polymer solution or aqueous dispersion on a carrier or another substrate.
- *Extrusion coating* is similar to film casting, but in this method the melt is deposited on the carrier or another substrate.

Finished films can be either laminated by several methods, or stretched (oriented) in machine direction, or cross-machine direction or biaxially. Stretched films are often annealed when desired. Other common treatments of finished polymeric films are surface treatment and metallization.

Typical uses of polymeric films include packaging, bags, labels, in automotive, aerospace, electrical and electronic, in chemical industry, optical industry, in the military, and in building construction and landscaping.

### 1.1.2 Fluoropolymers

Fluoropolymers are polymeric materials containing fluorine atoms in their chemical structures. Their chemistry is derived from the compounds used in the refrigeration industry, which has been in existence for more than 70 years [1]. The serendipitous discovery of PTFE in 1938 by Plunkett [2] in the laboratories of E.I. du Pont de Nemours and Company during the ongoing refrigerant research opened the field of fluoropolymers and their commercialization. PTFE was commercialized by that company as *Teflon* in 1950, but the technology had been used exclusively in the Manhattan Project already during the World War II [3].

Fluoropolymers in their simplest form are homopolymers or copolymers of saturated hydrocarbons in which all or some hydrogen atoms have been replaced by fluorine atoms or combination of fluorine and chlorine. Theoretically, there are many possible combinations but most commercially significant are derivatives of ethylene and propylene, sometimes referred to as fluorocarbon polymers. Other, more complex fluoropolymers are fluorinated acrylates, silicones, polyurethanes, polyimides, and other.

From general organic polymer concepts, there are two types of fluoropolymer materials: *perfluoropolymers* and *partially fluorinated polymers*. In the former case, all the hydrogen atoms in the analogous hydrocarbon polymer structures were replaced by fluorine atoms. Perfluoropolymers represent the largest volume of industrially processed fluorocarbon polymers, and the partially fluorinated polymers are mainly used in special applications. From the point of view of processing, the thermoplastic fluoropolymers, also referred to as fluoroplastics, there are melt-processible and nonmelt-processible fluoropolymers. Currently, the nonmelt-processible fluoropolymers include PTFE and PVF, while the remaining commercial fluoroplastics can be processed by melt-processing techniques.

The most common monomers used for the preparation of the known fluoropolymers are shown in Table 1.3. These can be combined to yield typically homopolymers, copolymers, and terpolymers. The resulting

**Table 1.3 Monomers Used in Commercial Fluoropolymers.**

| Compound | Formula |
|---|---|
| Ethylene | $CH_2 = CH_2$ |
| Tetrafluoroethylene | $CF_2 = CF_2$ |
| Chlorotrifluoroethylene | $CF_2 = CClF$ |
| Vinylidene fluoride | $CH_2 = CF_2$ |
| Vinyl fluoride | $CFH = CH_2$ |
| Propene | $CH_3CH = CH_2$ |
| Hexafluoropropene | $CF_3CF = CF_2$ |
| Perfluoromethylvinyl ether | $CF_3OCF = CF_2$ |
| Perfluoropropylvinyl ether | $CF_3CF_2CF_2OCF = CF_2$ |

products possess excellent properties, such as outstanding chemical resistance, weather stability, low surface energy, low coefficient of friction, and low dielectric constant. These properties come from the special electronic structure of the fluorine atom, the stable carbon–fluorine covalent bonding, and the unique intramolecular and intermolecular interactions between the fluorinated polymer segments and the main chains. Commercial fluorocarbon polymers are available in the two following distinct groups [4].

*Fluoroplastics*, which include a relatively large number of thermoplastic materials.

*Fluoroelastomers, which are used as raw materials for applications where elastomers with enhanced heat resistance, chemical resistance, resistance to aging and oxidation are required.* These are processed by techniques usual in the rubber industry and their typical applications include hose, belts, gaskets, seals, bladders, membranes, rubber-covered rolls, valve and pump linings, coatings, and sealants. It should be noted here that for the purpose of this publication, only fluoroplastics and to a degree fluorinated thermoplastic elastomers will be discussed since they are suitable for the use in commercial fluoropolymer films.

Due to their special chemical and physical properties, fluoropolymers are widely applied in the chemical, electrical/electronic, space, construction, architectural, food processing, medical and automotive industries, in engineering, and in military. Details are shown in Table 1.4. Table 1.5 lists major producers of thermoplastics.

**Table 1.4 Typical Applications for Fluoroplastics.**

| Industry | Applications |
|---|---|
| Aerospace | Wiring, special coatings, seals, flexible hose and tubing, foams in aircraft insulation, particularly for ducts and air-conditioning, electroluminescent lamps |
| Architectural | Roofing materials, architectural fabric, protective and decorative coatings |
| Automotive | O-rings, gaskets, shaft seals, head gaskets, fuel hose linings, flexible hose, valve stem seals, bearings and bushings, coatings, protective and decorative films, tubing, membranes in fuel cells |
| Chemical processing | Chemically resistant coatings and linings, pumps, pipe linings, impellers, tanks, heat exchangers, reaction vessels, autoclaves, valves and valve parts, flue duct expansion joints, solid pipes and fittings, bearings and bushings, sight glasses, flow meter tubes, flexible hose, filtration membranes, and textiles |
| Domestic | Nonstick coatings for cookware, nonstick utensils, cooking sheets for cookies |
| Electrical/electronics | Electrical insulation, wires and cables, flexible printed circuits, ultrapure components for semiconductor manufacture, condensers, batteries, barrier films for packaging of sensitive electronic parts, films for photovoltaic modules |
| Engineering | Bearings, bushings, gears, nonstick surfaces, low-friction surfaces, pipes and pipe coatings, fittings, valves and valve parts, seals and sealants, foams, conveyor belts for manufacture of ceramic tiles, outdoor signs |
| Food industry | Release sheets for fast food, conveyor belts for cooking and drying cookies and chips |
| Medical devices | Catheters, probes, cardiovascular grafts, heart patches, ligaments for knees, sutures, blood filters, tubings, dental floss, barrier packaging films for drugs |

**Table 1.5 Current Major Manufacturers of Fluoropolymers.**

| Manufacturer | Products |
|---|---|
| Arkema (www.arkema.com) | PCTFE, PVDF |
| Asahi Glass Co. (www.agc.co.jp) | ETFE, FEP, PFA, FEVE, PVDF, PTFE, Amorphous PTFE, PTFE Micropowders, Ionomers |
| Central Glass Co., Ltd. (www.cgco.co.jp) | Fluorocarbon TPE |
| Chemours (The Chemours Company) (www.chemours.com) | ETFE, FEP, PFA, PTFE, Amorphous PTFE, Modified PTFE, PTFE/PFA patented blends, PTFE Micropowders |
| Daikin Industries Ltd. (www.daikin.com), also Daikin America Inc. (www.daikin-america.com) | EFEP, ETFE, FEP, PCTFE, PFA, PTFE, PTFE Micropowders |
| DuPont USA (www.dupont.com) | Tedlar PVF and Tedlar S PVF films |
| Dyneon LLC (www.dyneon.com) | ETFE, FEP, THV Fluoroplastic, PTFE, Modified PTFE, PTFE Micropowders, PFA, PVDF |
| Honeywell (www.honeywell.com) | PCTFE |
| JSC Halogen (www.halogen.ru) | PTFE, ETFE |
| JSC Kirovo-Chepetsk Chemical Plant (www.kckk.ru) | ETFE, FEP, PVDF, PTFE, Modified PTFE |
| Kureha Chemical Industry Co. Ltd. (www.kureha.co.jp) | PVDF |
| Shandong Dongyue Chemical Co. (www.dongyuechem.com) | PTFE |
| Solvay Specialty Polymers (www.solvay.com) | ECTFE, MFA, PFA, PTFE, PVDF |
| Zaklady Azotove (www.azoty.tarnow.pl) | PTFE |

## 1.2 Manufacturing Methods for the Production of Films From Melt-Processible Thermoplastics

### 1.2.1 Film Extrusion

In general, most of the films made from melt-processible thermoplastics are produced by extrusion. Extrusion is a continuous process that involves forming a product (extrudate) in two dimensions. These $x-y$ dimensions determine the cross-sectional form of the extrudate and can be almost unlimited in scope, ranging from a simple tube to a very complex extruded profile. The third ($z$) dimension is the length of the extrudate. In principle, it can be infinite. In fact, it is limited by practical considerations governing winding, reeling, storage, and transport. The essential point is that extrusion always produces an object of constant cross section [5].

The product cross section is formed in a die; the extrusion process consists of heating a thermoplastic above its melt temperature and forcing it through the die. The heating and pressurizing device involves one or more screws operating in a heated barrel and is known as an *extruder*. In downstream of the die the extrudate is calibrated, cooled, and packaged by an array of ancillary devices including vacuum calibrators, air cooling chamber, water tanks, cooling rolls, haul-offs, cutters, and winders. In upstream of the die a melt pump (Fig. 1.1) may be interposed between the extruder and the die [6]. The exact selection and arrangement of these components of an extrusion system will depend on the end product. The specific processes for the production of different extruded products will be discussed in detail later in this chapter. The principle of extrusion is common to all these processes and will be discussed first.

The essential equipment for the extrusion process is the extruder, the function of which is to heat the plastic material to a homogeneous melt and to pump it through the extrusion die at a constant rate. Because the extrusion of plastics is a continuous process, the melt preparation device must be capable of a constant output. Thermoplastic extrusion depends almost entirely on the rotating screw as a melt delivery device. Thermoplastics are generally characterized by low thermal conductivity, high specific heat, and high melt viscosity, so the preparation of a

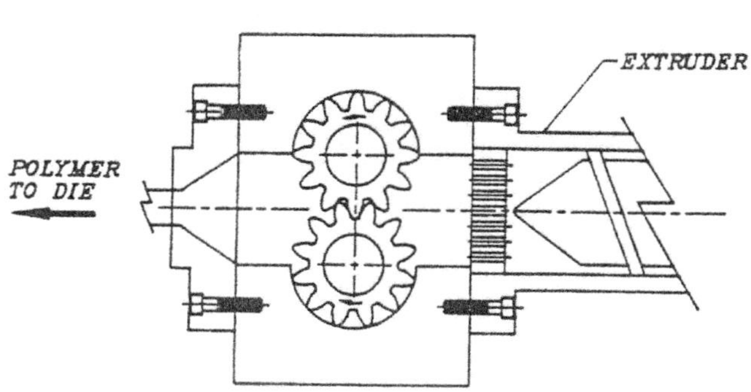


**Figure 1.1** Schematic diagram of a melt pump [6].

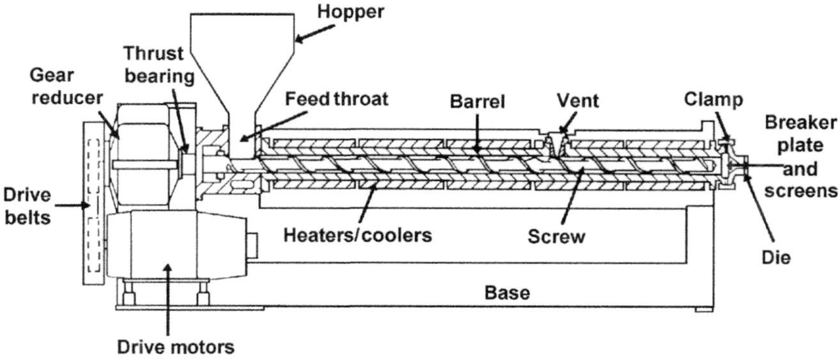


**Figure 1.2** Typical single-screw extruder with a vented barrel [5].

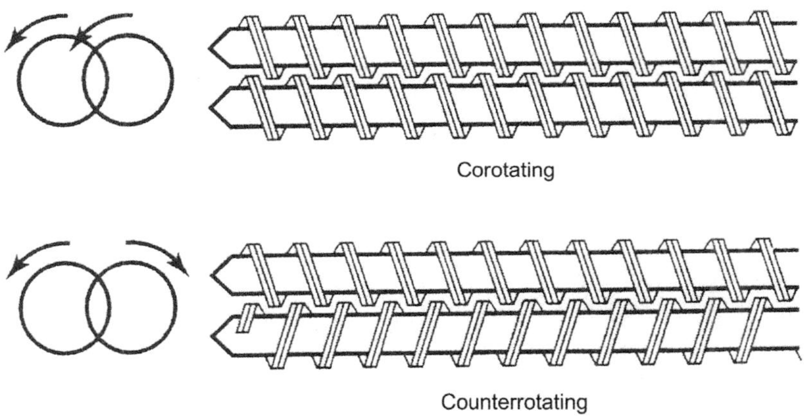


**Figure 1.3** Corotating and counterrotating twin screws [5].

uniform melt and its delivery at adequate pressure and a constant rate pose considerable challenges, which have been countered by the introduction of various types of extruders.

The principle variants are the single-screw and the twin-screw types. Of these, the single-screw extruder (Fig. 1.2) is by far the most popular. The twin-screw extruder may have parallel or conical screws that may rotate in the same (corotating) or in opposite directions (counterrotating) (see Fig. 1.3). Extruders with more than two screws are known, for example, the quad-screw extruder, but they are not widely used [5]. Twin-screw extruders are normally used when mixing and homogenization of the melt is very important, and in particular where additives are to be incorporated. Fluoropolymers are normally extruded using a single-screw machine.

The single-screw extruder consists essentially of a screw that rotates in an axially fixed position within the close-fitting bore of a barrel. The screw is motor-driven through a gear reduction train and is supported by a thrust bearing that opposes the force exerted on the plastic melts. A helical flight on the screw provides the drag-induced conveying motion that develops this force. The barrel is equipped with both heating and cooling means, and its downstream end is provided with an attachment device for a shaping die that determines the cross section of the extruded product. The upstream or inlet end of the barrel is equipped with a feed throat or an aperture in the barrel wall where a plastic material is input, generally in the form of granules or pellets. During its passage along the helical screw flight, this material is heated by a combination of conducted heat received from the barrel and mechanical shear heat derived from the mixing and kneading action of the screw. The output rate of the extruder is a function of screw speed, screw geometry, and melt viscosity. The pressure developed in the extruder system is a function of melt viscosity, screw design, and barrel and die resistance. Extrusion pressures are lower than those encountered in injection molding and typically less than 35 MPa (5000 psi) [6].

The key determinant of extruder performance is the screw. The screw has three functions to perform: feeding and conveying the solid thermoplastic feed (most frequently pellets); melting, compressing, and homogenizing the material; and metering or pumping the melt to the die. The typical extruder screw (Fig. 1.4) takes the form of a single constant-pitch flight that decreases in depth from the input end to the output end. The pitch is usually equal to the screw diameter.

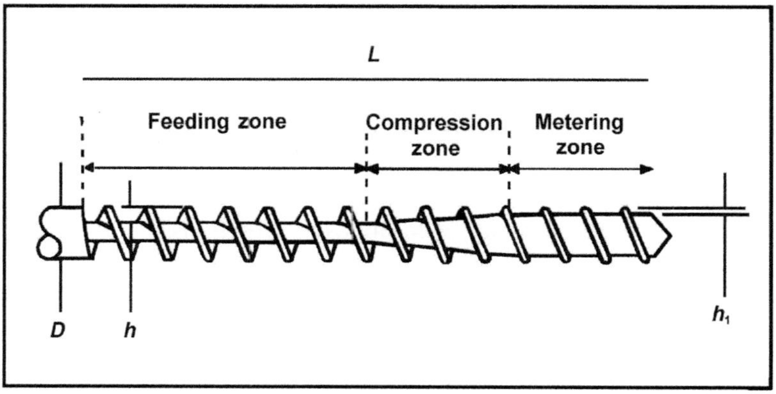


**Figure 1.4** Features of a typical extrusion screw [6]. *D*, Diameter; *h*, initial length; $h/h_1$, compression ratio; $h_1$, final flight depth; *L*, screw length.

This is sometimes known as a square-pitch screw; the resulting helix angle is 17.8 degrees. The screw features three sequential zones, corresponding to the three functions of feeding, compression, and metering.

Flight depth is usually constant in the feeding and metering zones and decreases at a constant rate over the compression zone. The feed zone occupies about 50% of the screw length; the compression zone takes up 25%–30%, and the metering zone accounts for the balance. The ratio of the flight depths in the feeding and metering zones is known as the compression ratio, and it affects the mixing and shear-heating characteristic of the screw. The ratio of the screw length to its diameter is known as the *L/D ratio*, which has a bearing on mixing and uniformity of output. Most conventional screw designs perform satisfactorily for thermoplastic elastomers. Extrusion screws for wire insulation should have an L/D ratio in the range of 24:1–30:1. A screw with a long feed section with approximately a 3:1 compression ratio (core progressive profiles that do not subject the resin to excessive shear) is preferable.

The mixing performance of a single screw can be improved by the addition of mixing elements in the metering zone. These are sometimes teamed with or replaced by static elements such as pins in the barrel. Attention to the barrel can also improve the material conveying performance at the inlet, stepping up the output rate. A series of axial grooves in the barrel wall (Fig. 1.5), extending for at least three screw diameters, is effective.

Variants on the single screw include the *barrier* or *melt extraction screw* (Fig. 1.6) and the *vented screw*. Barrier screws have separate input and output flights that overlap in the midsection of the screw

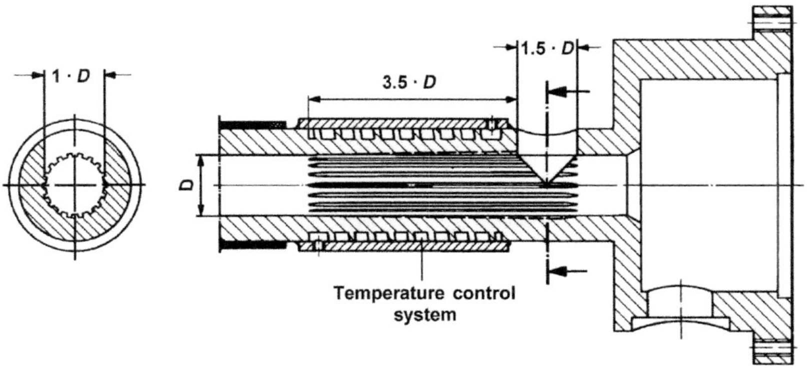


**Figure 1.5** Grooved feed section of barrel [6].

**Figure 1.6** Section of a barrier screw [6].

where the plastic material is only partially melted. The barrier flights are slightly smaller in diameter than the main flights. Molten material passes through this clearance and proceeds to the metering zone, while solids are retained for further heating. The barrier screw offers the potential for lower melt temperatures or higher output rates but is difficult to design and optimize. Vented screws, sometimes known as decompression screws or two-stage screws, are effectively two screws placed in series, with the second or downstream screw having a higher transport rate than the first. The upstream end of the screw is provided with conventional feeding, compression, and metering zones, but these are followed by a sudden increase in flight depth, and then by further compression and metering zones. The mid-screw increase in flight depth causes a sudden pressure drop in the molten plastic, which allows any dissolved volatile matter to boil out. The lack of compression at this point, coupled with the differential transport rates of the two screw sections, makes it possible to extract the volatile material through a port in the extruder barrel without the plastics melt emerging at the same time. Vented screws typically have an L/D ratio of greater than 30:1 and suffer from some instability in pumping output. Operating conditions are also constrained by the need to ensure that the vent does not plug with plastics melt. For these reasons the use of vented screws is usually confined to materials likely to contain moisture, volatile matter, or entrained air.

## 1.2.2 Film Extrusion Methods

Films from thermoplastics can be produced either by extrusion casting or extrusion blowing processes. Each has its advantages and disadvantages. These basic processes result in film with a molecular orientation predominantly in the machine direction. Regardless of process, film production lines include common downstream equipment such as haul-off, tensioning, and reeling stations. A high-purity melt, free of inclusions, is essential for film production. This is achieved by filtering the melt through a *screen pack* located upstream of the die.

#### *1.2.2.1 Cast Film Extrusion*

Cast film is produced by extruding the melt from a slit die and cooling it either by contact with a chill roll or by quenching in a water bath. Both processes are characterized by relatively high melt temperatures and rapid rates of film cooling. This results in films with low haze, good clarity, and high gloss. Cast film grades typically have a melt flow index in the range of 5.0–12.0 g/10 minutes.

In the chill-roll cast film process, a plastics web is extruded from a slit die (Fig. 1.7) against the surface of a water-cooled chill roll [7]. The die is arranged to extrude vertically or obliquely downward so that the film web is delivered approximately tangentially to the roll surface. The die is similar in principle to a sheet die but will usually not include a restrictor bar. Film thickness is partially regulated not only by the gap between the die lips but also by the rotational speed of the chill roll, which is arranged so as to draw down and thin the melt web. Consequently, the die gap is set in excess of the desired film thickness. Typical die gap settings for most thermoplastics are 0.4 mm for films up to 0.25 mm (0.01 in.) thick, and 0.75 mm (0.03 in.) for film gauges in the range of 0.25–0.6 mm (0.01–0.024 in.). Die lip adjusters should allow the die gap to be varied at each adjustment point across the die width in order to allow control of the transverse film thickness. The quality of film reel will suffer if the transverse thickness tolerance exceeds ±5% of target thickness. A constant temperature should be maintained across the die so that film drawdown rates and physical properties remain constant across the film web. Attempts to control film thickness by varying the temperature profile across the die will disturb these factors and reduce film quality. When the process is correctly regulated, the thickness uniformity of chill-roll cast film is substantially superior to blown film.

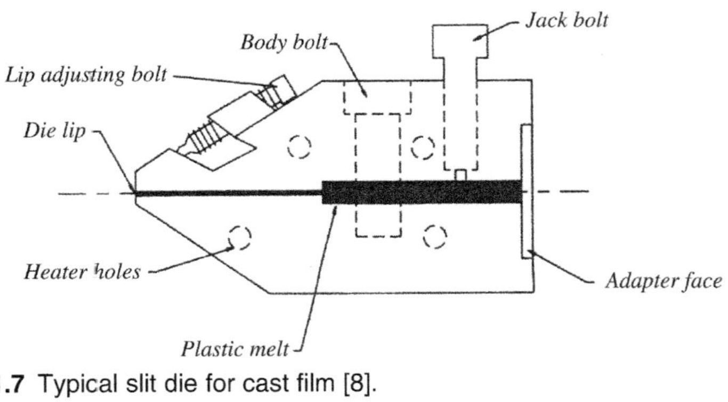


**Figure 1.7** Typical slit die for cast film [8].

The die is maintained in close proximity (typically 40–80 mm or 1.6–3.2 in.) to the chill roll so that the low-strength melt web remains unsupported for a minimal distance and time (Fig. 1.8). If the die is too close, there is insufficient space for thickness drawdown and widthwise neck-in to take place in a stable manner. The web flows on to the water-cooled chill roll with a wraparound of 240 degrees or more before passing to a second chill roll and then proceeding to edge trimming, tensioning, and windup stations. The first chill roll is critical to process quality. The cooling capacity must be sufficient to chill the film at high output rates, and the temperature gradient across the width of the roll should not exceed ± 1°C (1.8°F). The actual roll temperature depends on film gauge, line speed, and roll diameter. The chill-roll drive speeds must be very precisely regulated to control film drawdown and finished thickness. At line speeds greater than 30 m/min (98 ft./min), there is a tendency for a thin cushion of air to become trapped between the film and the chill roll. This results in slow and uneven cooling, affecting the appearance and properties of the film. Two measures are adopted to counteract the problem (Fig. 1.8). An air knife delivers a streamlined jet of filtered air that impinges on the film just beyond the point of first contact with the chill roll and presses it against the roll. The air is supplied through a narrow slit of about 1.5 mm (0.06 in.) gap and controlled at a low differential above atmospheric pressure. An optimum air knife setting improves film clarity and gloss, but excessive pressure induces melt vibration and mars the surface of the film. The second measure is the provision of a vacuum box, associated with the die, and operating close to the chill-roll surface just ahead

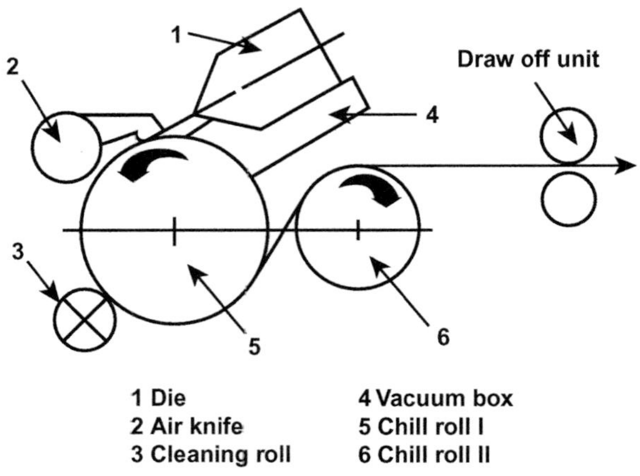


**Figure 1.8** Detail of chill-roll process [7].

of the extruded melt web. The vacuum box removes any condensates that have been deposited on the roll by the film and tends to draw air out of the interface between film and roll. The film temperature should be close to ambient at the film windup station; otherwise the reel will continue to shrink after cooling, causing stretch marks and corrugations, and accentuating any variations in thickness. Winding tension should be sufficient to ensure the integrity of the reel but otherwise should be kept low to allow for a small amount of postwinding shrinkage that will tighten the reel. A typical line for chill-roll cast film is in Fig. 1.9.

As in other plastic processes, the cast film process depends on many interactive variables, so that any defect may have one of several causes. A remedy for one defect may introduce another, so process problem-solving is not straightforward.

The *water-quench cast film process* (Fig. 1.10) is similar in concept to the chill-roll process and uses similar downstream equipment. A water bath takes the place of the chill roll for film cooling, and by cooling both sides of the film equally, it produces a film with slightly different properties compared to chill-roll cast film. The extruder's slit die is arranged vertically and extrudes a melt web directly into the water bath at close range. The film passes under a pair of idler rollers in the bath and, for any given rate of extrusion, it is the rate of downstream haul-off that regulates film drawdown and finished thickness. The speed of the process is limited by the tendency of the film to carry over water from the quench bath. Surface defects can also arise from rippling in the water bath. The very rapid quenching induced by the water bath reduces crystallinity and produces a tough film.

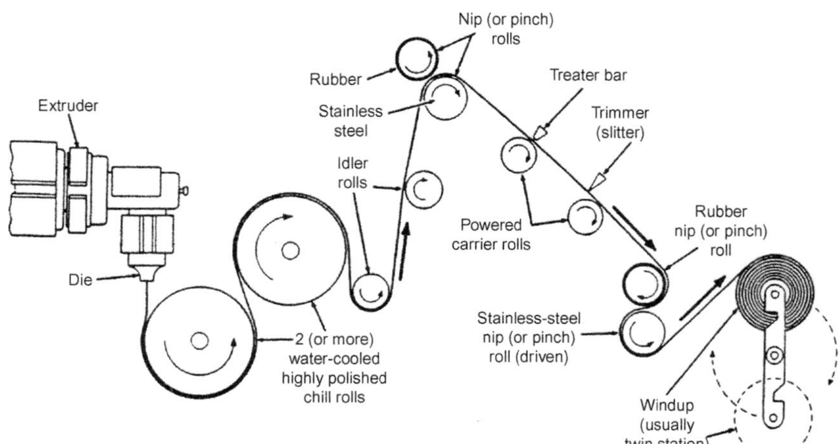


**Figure 1.9** Typical line for chill-roll cast film [7].

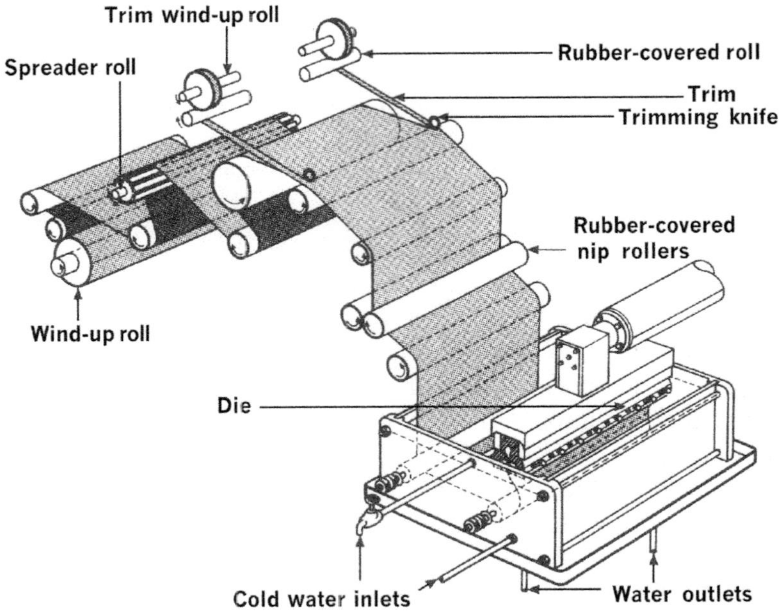


**Figure 1.10** Typical water-quench film line [7].

### *1.2.2.2 Blown Film Extrusion*

The blown film process involves extruding a relatively thick tube, which is then expanded or blown by internal air pressure to produce a relatively thin film. The air-cooled blown film process is in very widespread use mainly for polyolefin films. The process needs a melt strength greater than that required by cast film processes so lower melt temperatures are used. The die should be designed for constant output rates and thickness at every point around the annular die gap. This requires streamlined internal melt flow paths and precise multipoint means of centering the mandrel within the die ring. The die gap is typically about 0.4 mm (0.015 in.), with a short land length and a die entry angle of about 10 degrees. Close thickness tolerances are difficult to achieve with the blown film process; variations around the bubble can be limited to perhaps $\pm$ 10% of the target film thickness. To distribute these variations evenly, blown film lines may include rotating or oscillating motions applied either to the die, the extruder, or the haul-off.

As it emerges from the die, the tube passes through an air ring that lays a large volume of low-velocity cooling air over the external surface. At the same time the tube is pressurized internally by air supplied

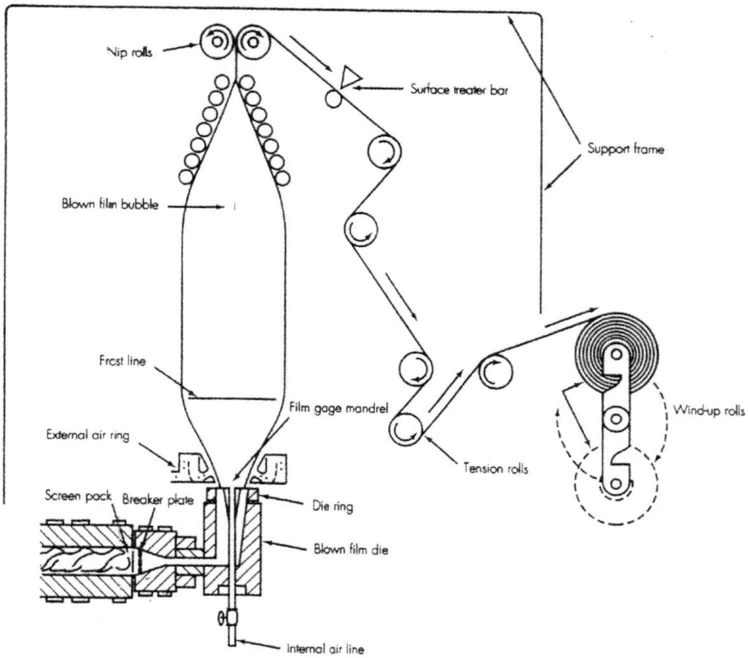


**Figure 1.11** Air-cooled blown film process [7].

through the die mandrel. The air is confined by downstream nip rollers, so it inflates the still soft tube to form an enlarged bubble. The amount that the bubble is extended is called the *blow-up ratio* (BUR) and is typically in the range of 2:1–4:1. The distance that the tube travels upward before its diameter increases is called the *stalk height*. Most materials are blown with a short stalk, with the bubble diameter being increased to the final size shortly after leaving the die lips [8].

The line speed is effectively limited by the bubble length available for cooling. Downstream equipment includes bubble calibration, a collapsing frame, and haul-off, tensioning, and reeling stations. Film thickness and process control is a balance between cooling rate, bubble length, BUR, and film tension.

After the bubble is cooled and collapsed into a flat shape, the product (layflat) is rolled up on a high-speed winder. In some cases the layflat is slit to form two separate sheets and wound onto individual rolls, requiring two independent winding shafts (Fig. 1.11).

The *water-quench blown film* process is also known as the tubular water-quench process and uses a tube die that is arranged to extrude vertically downward (Fig. 1.12). The bubble size is limited and

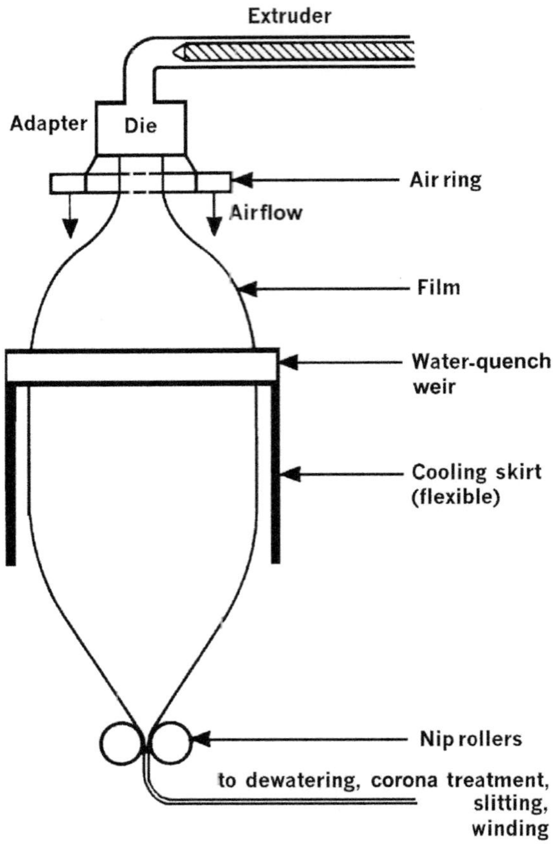


**Figure 1.12** Water-quench process for blown film [6].

calibrated by an annular water cooling weir that covers the outer bubble surface with a stream of cooling water. The water contact is directed and controlled by an annular flexible skirt [9]. The rapid bubble cooling induced by the water stream limits crystallinity and produces a clearer film than is possible by the air-cooled blown process. After cooling the film passes through a collapsing frame and proceeds through dewatering, haul-off, and winding stations.

### *1.2.2.3 Coextrusion*

Coextrusion is the process of forming an extrudate composed of more than one thermoplastic melt stream. The process came about because some service demands, particularly from the packaging industry, could not be satisfied by a single polymer although they could be met by a combination of polymers. Coextrusion was first practiced in the production of cast film and is now also used in blown film and sheet extrusion. The intention is normally to produce a laminar structure in which each layer contributes a key property to the overall product performance. Coextruded films may be very complex structures composed of many different functional layers, including tie layers whose purpose is to bond neighboring layers of limited compatibility. Five layers are not uncommon. However, side-by-side coextrusion is also possible.

Separate extruders are required for each distinct material in the coextrusion. The process has two variations, depending on the point at which the separate melt streams are brought together. In feed block coextrusion the streams are merged into a single laminar melt flow in a feed block (Fig. 1.13), which is positioned immediately upstream of the extrusion die. The process depends on the high viscosity of plastics melts to prevent intermingling of the layers as they pass through the extrusion die. The flow rate of each component layer can be controlled by valves in the feed block and the capital cost is relatively low. The alternative, die coextrusion, uses a complex die construction in which separate melt path manifolds are arranged to merge at a point close to the die exit (Fig. 1.14). The thickness and flow rate of individual layers can be independently controlled and also possible to handle polymers with substantially differing viscosities and melt temperatures. The capital and maintenance costs of such multimanifold coextrusion dies are high.

Coextrusion of blown film is accomplished by multimanifold dies. The distribution of melt along the circumference of these dies is accomplished by a spiral mandrel [10]. An example of a die for the coextrusion of a three-layer film is in Fig. 1.15 [8], and an arrangement for the coextrusion of a nine-layer blown film is in Fig. 1.16.

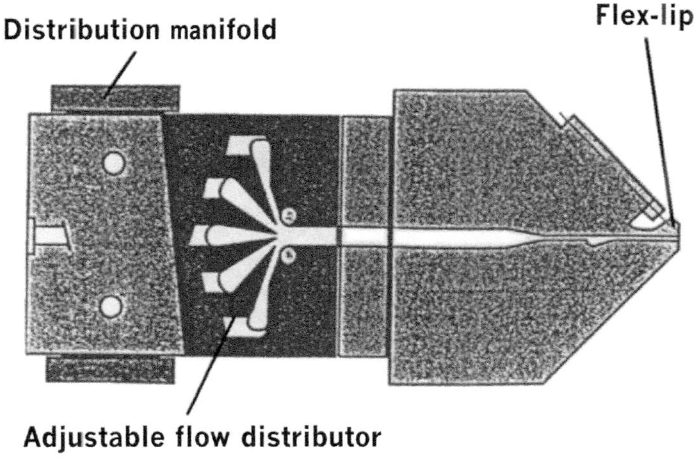


**Figure 1.13** Schematic of coextrusion feed block [9].

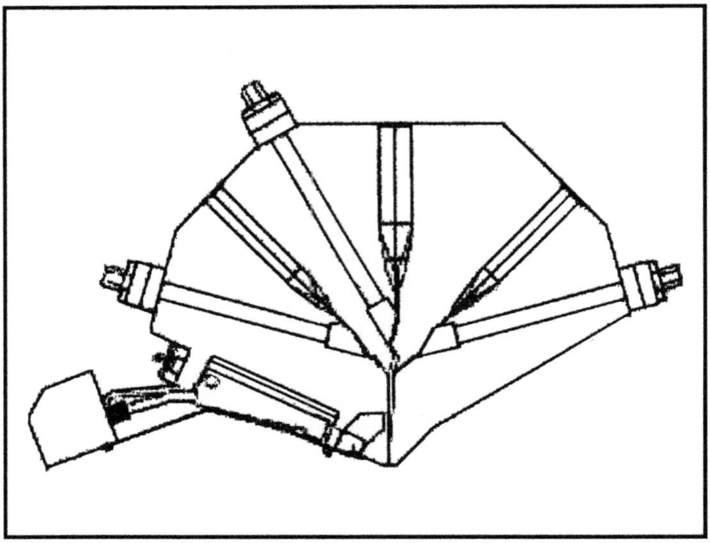

**Figure 1.14** Three-layer multimanifold coextrusion die [9].

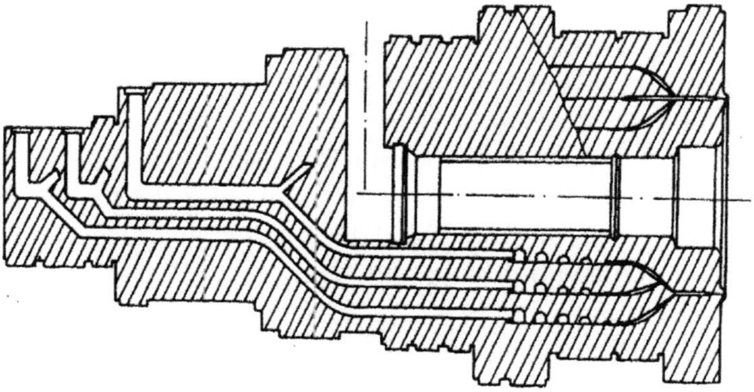

**Figure 1.15** Three-layer coextrusion die for blown film. Courtesy DPA.

## 1.3 Secondary Processing of Thermoplastic Films

Most of the films produced commercially are used without further operations other than slitting. However, in order to expand their utility, they have to be subjected to additional operations, such lamination with other films, textiles, rubber, metals, and other materials or bonded to the surface of structures. For these purposes, they need to have their surface modified to assure satisfactory bonding to the surface of other

**Figure 1.16** Equipment for coextrusion of nine-layer blown film. Courtesy Battenfeld Gloucester Engineering Company.

substrate. Some of the commonly used methods for that are discussed in Section 1.3.1. Additional processes involved are heat sealing, metallization, and orientation (stretching). These are covered in the remainder of this chapter.

### 1.3.1 Surface Preparation of Thermoplastic Films

Since adhesive bonding is a surface phenomenon, preparation of surface prior to bonding is very important. Both surfaces bonded together (adherends) have to be made receptive to the development of a strong, durable adhesive bond. Proper surface preparation should ensure that the bond failure occurs within the adhesive layer, not at the surface. This ideal failure is called cohesive failure. Bond failure at the surface of the adherends is called adhesive failure [11]. The simplest way to assure adhesive is to have both surfaces clean. Metals and most nonorganic adherends have to be free of oils, greases, rust, and other foreign matters. This is accomplished by washing by solvents, sanding, and grit blasting. *Surface energy* is the property that determines the surface wetting by adhesive or compatibility of two melted adherends. Plastics being organic materials have inherently lower surface energy than metals and other solids and tend to form intrinsically poor adhesive bonds without any type of surface treatment. There are many physical and chemical treatments used for that purpose. Industrial surface treatments for plastics include corona, flame and plasma treatment, and chemical etching [12]. The increase in surface energy of plastics occurs through the surface oxidation of the polymer chains and for halogenated polymers, such as PVC and fluoropolymers, surface modification involves significant dehalogenation, the removal of chlorine and/or fluorine atoms from the surface of the macromolecules [13].

#### *1.3.1.1 Corona Treatment*

Corona discharge takes place at atmospheric pressure in contrast to low-temperature (or cold) plasma, which requires vacuum. Corona is a stream of charged particles that are accelerated by an electric field. It is generated when a space gap filled with air or other gases is subjected to sufficiently high voltage in order to set up a chain reaction of high-velocity particle collision with neutral molecules, resulting in the generation of more ions [14]. The concept is illustrated by Fig. 1.17. The conventional corona treatment system is in Fig. 1.18. In this method the

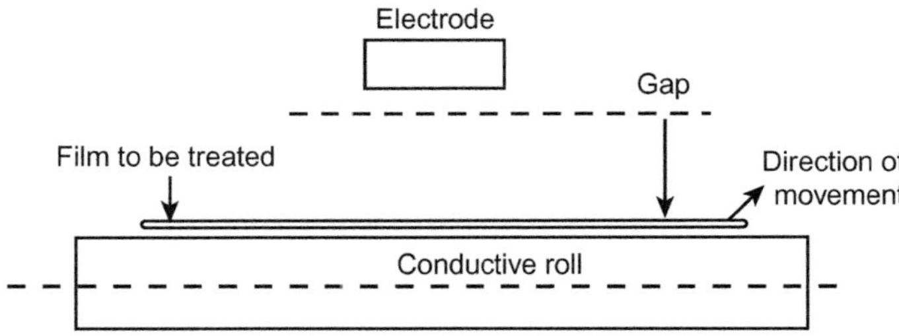


**Figure 1.17** Conceptual schematic diagram of a film corona treatment [14].

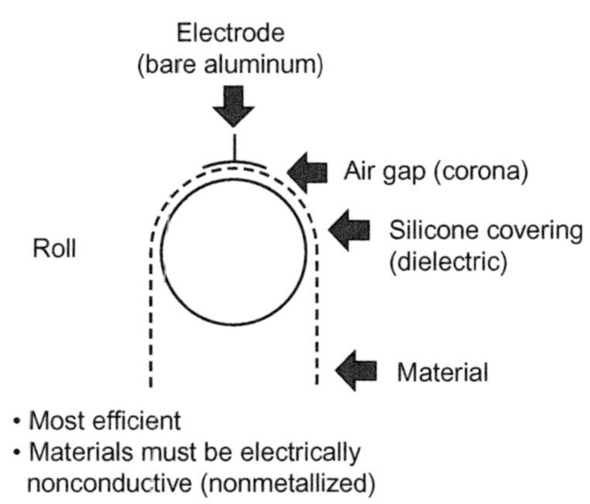


**Figure 1.18** Conventional configuration of corona treatment equipment [15].

plastic article is exposed to a corona produced by high-frequency, high-voltage alternating current. Frequently, corona treatment equipment is attached to the film extrusion equipment. It is reported that corona treatment is used for FEP and PTFE films [16]. Corona treatment is believed to roughen the plastics by the degradation of the amorphous regions of the polymer without affecting crystalline regions. The ensuing roughness of the surface provides a larger adhesive contact area than a smooth surface [17].

#### 1.3.1.2 Plasma Treatment

Plasma is considered the fourth state of the matter and is produced by exciting a gas with electrical energy introduced into a vacuum chamber. It is a collection of charged particles, containing positive and negative ions with other types of fragments, such as free radicals, atoms, and molecules possibly present. Since plasma is intensely reactive, it can effectively modify surfaces of plastics. It can be used to treat plastics including films to impart hardness, roughness, more or less wettability, and increased adherability to the part surfaces. Plasma treatment oxidizes the surface of the polymer in the presence of oxygen, which is believed to be the reason for roughening of the surface.

*Atmospheric plasma treatment* (APT), also called *glow discharge*, is operating without the use of vacuum. The plasma creates uniform plasma cloud that completely surrounds small objects or spreads into the boundary layer of the surface. Both films and three-dimensional parts can be treated by using APT technology. Fluoropolymer films can be treated by APT with PVF responding well to it but PTFE does not [18].

It should be noted that there are several companies offering in-house contract for full-service plasma treatment.

#### 1.3.1.3 Flame Treatment

Flame treatment is defined as a surface preparation technique in which the plastic is briefly exposed to a flame. Flame treatment oxidizes the surface through a free radical mechanism, introducing hydroxyl, carbonyl, and amide functional group to depth of about 4–6 nm, and produces chain scissions and some cross-linking. It is a commercial process that increases wettability and interfacial diffusivity. The film is passed over an oxidizing flame formed by oxygen-rich (relative to stoichiometry) mixture of hydrocarbon gases. However, this method is not effective in the adhesion treatment of perfluoroplastics [19].

#### *1.3.1.4 Chemical Etching*

Chemical etching is the most effective method of entering the surface of plastics optimal for bonding. This technique can alter the physical and chemical properties of the surface. Chemical treatment is often preceded by a cleaning operation to remove surface contamination. Typically the surface of the plastic part is washed by soap and water or an organic solvent followed by immersion in the chemical treatment bath. The solution contains an acid or base, oxidizing agent, chlorinate agent, or other active agents. Afterward the part is rinsed in water followed by drying at elevated temperature.

#### *1.3.1.5 Sodium Etching of Fluoroplastics*

Perfluorinated fluoroplastics are chemically unaffected by nearly all commercial chemicals with the exception of highly oxidizing substances as elemental forms of sodium, potassium, and other alkaline metals. This is the basis for sodium etching of fluoroplastic parts. The original method for surface treatment for adhesive bonding was developed for PTFE and was etching by a sodium solution in anhydrous liquid ammonia. An alternative solution of sodium is to prepare a complex with naphthalene followed by dissolution in tetrahydrofuran or dimethyl glycol ether. Newer systems using glycol diethers (referred to as *Glymes*), which are much less toxic than tetrahydrofuran.

Special precautions must be taken while working with sodium etching solutions. Fluoropolymer products (films and parts), treated by sodium etching solution, should be stored in cold dark atmosphere free from oxygen and moisture. The useful shelf life of etched polymer under these conditions at temperature lower than 5°C is 3–4 months. This subject is covered thoroughly in Ref. [20].

For in-house etching, it is advisable to consider purchasing commercial etching solutions available from a number of sources [21]. Some of them can be used successfully for fluoropolymers other than PTFE. Moreover, there are several companies offering full-service contract etching [22].

### 1.3.2 Lamination of Thermoplastic Films

Lamination is the technique of manufacturing a material in multiple layers, so that the composite material achieves improved strength, stability, sound insulation, appearance, or other properties from the use of differing materials. A laminate is a permanently assembled object by heat, pressure, welding, or adhesive [23].

In film lamination a fabricated film is adhered to a moving film or substrate mostly by the application of heat and pressure. Film lamination methods include hot roll, belt, flame, calander lamination, and sheet extrusion; each type providing a different combination of heat and pressure.

#### 1.3.2.1 Hot Roll/Belt Lamination

Hot roll and belt lamination use heat and pressure as their means of bonding. Usually one of the films has been coated with heat-activated adhesives applied to it in precedent process. As shown in Fig. 1.19, the two films are drawn onto heated rollers where the materials are heated and pressed together. The heat starts melting the adhesive and activates it, creating a bond when pressed against the opposite material. Film lamination can be also done by using calander with heated rolls [24] (see Fig. 1.20). Another arrangement is extrusion lamination, in which case one of the films is produced by extrusion and is laminated to the other substrate (film) by hot roll or calander [24].

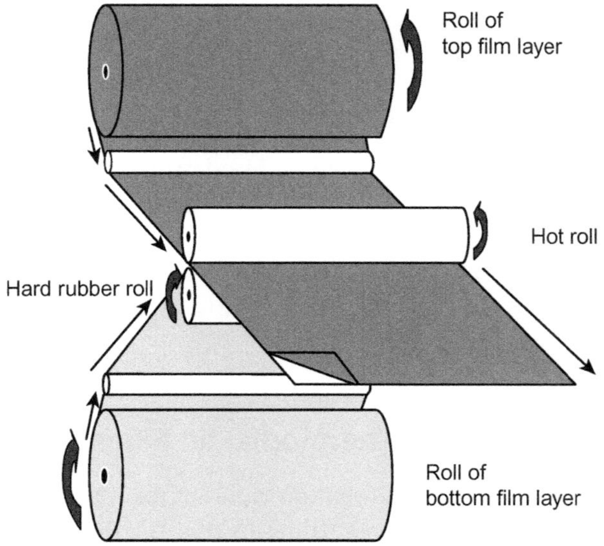


**Figure 1.19** Schematic of a hot roll laminator [23].

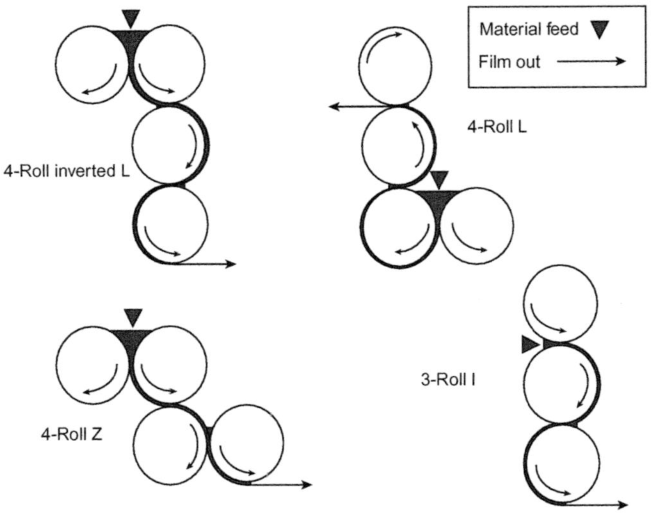


**Figure 1.20** Schematic of three common calander roll configurations [24].

Double belt lamination is similar to the hot roll lamination. The equipment for that typically include top and bottom belts with heating elements and pressure mechanisms that transfer heat and pressure to the products between the belts. One of the major advantages to this machine is its ability to continuously hold product under heat for lamination and then for cool to set good strength. Product is held securely throughout complete process [25].

#### 1.3.2.2 *Flame Lamination*

Flame lamination is often used to bond film and/or fabric to soft polyurethane foam. The process involves passing of the soft foam over an open fire, which creates a thin layer of molten polymer. In this case the molten polymer acts as adhesive. The film and/or fabric is quickly pressed against the foam while it is still in the molten state [23].

### 1.3.3 Heat Sealing of Thermoplastic Films

Heat sealing is a widely used process in the packaging industry. It involves joining two polymer films by the application of heat and pressure for a specified time. The polymer characteristics that play a critical role in the heat sealing process involve melting temperature, chain diffusion rate, melt strength, and crystallization kinetics [26]. Under the instantaneous application of heat and pressure, the surfaces of the film come in intimate contact and melt, which leads to diffusion of chains across the interface forming molecular entanglements [26–28]. Thermoplastic films, including fluoroplastic films can be sealed by any method that heats the contacting surfaces of the film above the melt temperature of the given polymer and at the same time. Applied pressure provides intimate contact of those surfaces. In case the molten film tends to stick to the jaws of the metal sealer, it is necessary to use a release agent. Polyimide film (e.g., Kapton) may also be used as a release slip sheet. The usual techniques used are *hot bar heat sealing*, *impulse heat sealing*, and *hot air sealing* [29].

### 1.3.4 Metallization of Thermoplastic Films

Plastic parts can be coated with metal, in a process called metallization, for both aesthetic and mechanical purposes. Visually, a metal-coated piece of plastic features increased gloss and reflectivity. Other properties, such as abrasion resistance and electric conductivity, which are not innate characteristics of plastic, are often obtained through metallization. Metalized plastic components are used in similar applications as metal plated parts but tend to be lower in weight and have higher corrosion resistance, although not in all cases. In addition, electrical conductivity can be controlled in metalized plastic components, which are inexpensive to manufacture. To metalize a piece of plastic, several common methods are used: arc and flame spraying, and plating and vacuum metallization [30]. It is also possible to metalize a transfer film and use alternative methods to apply the film to the surface of the substrate.

#### *1.3.4.1 Arc and Flame Spraying*

In basic flame spraying a handheld device is used to spray a layer of metallic coating on the substrate. In flame spraying the primary force behind deposition is a combustion flame, driven by oxygen and gas. Metallic powder is heated and melted, as combustion flame accelerates the mixture and releases it as spray. This process has a high deposition rate and creates very thick layers, but the coatings tend to be porous and somewhat rough. Due to the nature of the application process, coatings can be applied to specific areas of components, which is useful when working with complex or unusually shaped components. The process is relatively easy and requires minimal training.

Arc spraying is similar to flame spraying, but the power source is different. Instead of depending on a combustion flame, arc spraying derives its energy from an electric arc. Two wires, composed of the metallic coating material and carrying DC electric current, touch together at their tips—the energy that releases when the two wires touch heats and melts the wire, while a stream of gas deposits the molten metal onto the surface of the substrate, creating a metal coat. Like flame spraying, the resulting coating typically suffers from high porosity.

#### *1.3.4.2 Electroless Plating and Electroplating*

Plating is typically divided into two categories, depending on the presence of electric current. In electroless plating [30], electric current is not used, whereas in electroplating, electric current is used. Both processes tend to be more effective than vacuum metallization at producing metallic coats with strong adhesion, although plating tends to be more dangerous.

*Electroless plating* is often used to deposit nickel or copper metal onto plastic substrates. The surface becomes susceptible to hydrogen bonding as a result of the contact with the oxidizing solution and this improves the bonding of the metallic coating [31]. Coating occurs when the plastic component (postetching) is immersed in a solution containing metallic (nickel or copper) ions, which then bond to the plastic surface as a metallic coating. Usually the thickness of electroless metal coat is 0.5 mil (13 μm).

In order for *electroplating* (or *electrolytic plating*) to be successful, the plastic surface must first be rendered conductive, which can be achieved through basic electroless plating [30]. Once the plastic surface is conductive, the substrate is immersed in a solution. In the solution are metallic salts, connected to a positive source of current (cathode). An anodic (negatively charged) conductor is also placed in the bath, which creates an electrical circuit in conjunction with the positively charged salts. The metallic salts are electrically attracted to the substrate, where they create a metallic coat. As this process happens, the anodic conductor, typically made of the same type of metal as the metallic salts, dissolves into the solution and replaces the source of metallic salts, which is depleted during deposition.

#### *1.3.4.3 Vacuum Metallization*

Vacuum metallization also called physical vapor deposition is a process of depositing metal coatings on prepared surfaces. Before the process can begin the plastic component is washed and coated with a base coat and/or etched, so that the metal layer is smooth and uniform. Next, a metal (typically aluminum) is evaporated in a vacuum chamber [32]. The evaporation of the metal takes place by feeding aluminum wire onto heated sources or "boats," which operate at approximately 1500°C (2700°F). The vapor then condenses onto the surface of the substrate, leaving a thin layer of metal coating. The entire process takes place within a vacuum chamber to prevent oxidation. The deposited metal layer is very thin (typically 0.01−0.025 μm) and increases mainly the barrier properties of the substrate with respect to water, gases, and light [33] and the number of other applications in electronics, aerospace, etc.

Fluoropolymer films (PTFE, FEP, ETFE, ECTFE, PVDF, and PVF) can be readily metallized by a variety of metals by vacuum and electroless deposition [29,34,35]. Aluminum-coated PTFE is used in high-temperature-resistant capacitors [36].

### 1.3.5 Orientation

The process of making films usually results in the polymer chains arranging in a relatively random order thus exhibiting anisotropy. If the film is stretched, the polymer chains tend to line up (or orient) in the direction of the stretch. Consequently, this orientation affects the physical properties of the film in the direction of the stretch as follows [24]:

- increased tensile properties, better resistance to break and tear;
- improved optical properties, such as improvement in clarity, reduction of haze, and increased surface gloss;
- controlled shrinkage;
- improved barrier properties, particularly to water and oxygen;
- increased toughness and puncture resistance; and
- increased stiffness.

Orientation in one direction is very common, and there are essentially two types:

1. *machine direction orientation* (MDO) that is stretching the film in the direction of the process flow and
2. *transverse direction orientation* (TDO) that is stretching the film across the process flow.

Another possibility is the stretching in both directions, which is referred to as *biaxial orientation* abbreviated as BO.

In industrial practice, there are specific machines and corresponding processes for these tasks, which are as follows.

#### 1.3.5.1 Machine Direction Orientation

MDO is accomplished by passing the film through a series of heated rolls rotating at controlled and increasing speeds [24]. The degree of stretch is controlled by varying the roll speeds. The process consists of the following steps:

1. *Preheating* to soften the film and allowing its stretching without breaking.
2. *Drawing* that stretches the film (up to 10 times) and forces alignment of the polymer molecules in the film. In this section, one drawing roll runs faster than the other.
3. *Annealing* locking in the properties and controlling shrinkage.
4. *Cooling*.

Schematic of the equipment for MDO is in Fig. 1.21 [37].

#### 1.3.5.2 Transverse Direction Orientation

TDO takes place in the direction that is 90 degrees to the machine direction in a machine called *tenter frame* [24,38,39]. Tentering equipment consists essentially of a temperature-regulated tunnel in which the film edges are gripped by chain-driven tension clips running in divergent paths (see Fig. 1.22) [39]. The tenter clips close onto the film edge at the entrance of the tenter frame and move it forward while stretching it gradually over the width as they move further apart. Several types of tenter frame clips are shown in Fig. 1.23. Transverse stretch is

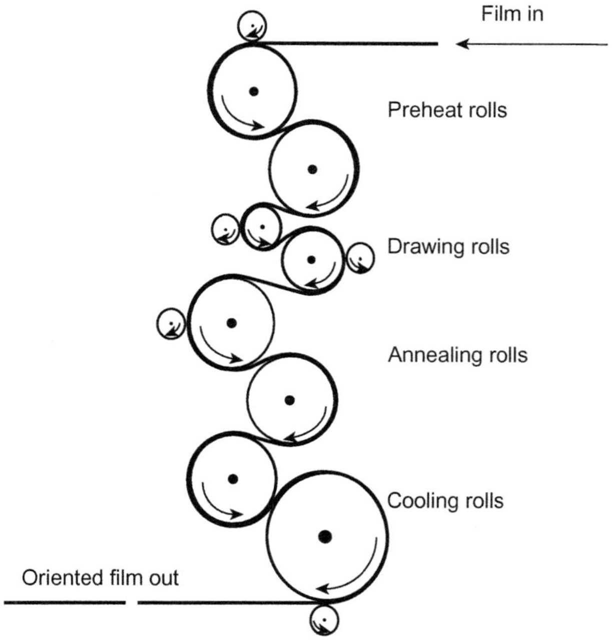


**Figure 1.21** Schematic of a system for machine direction orientation [24].

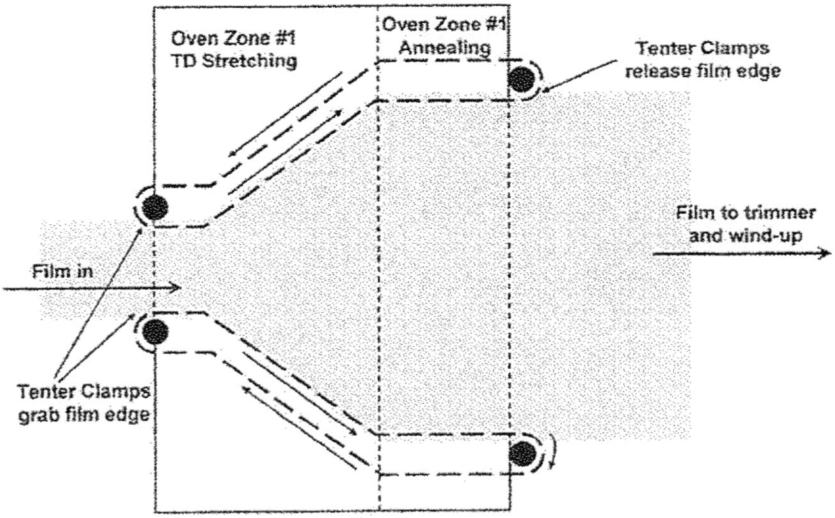


**Figure 1.22** Schematic of a tenter frame for transverse direction orientation [39].

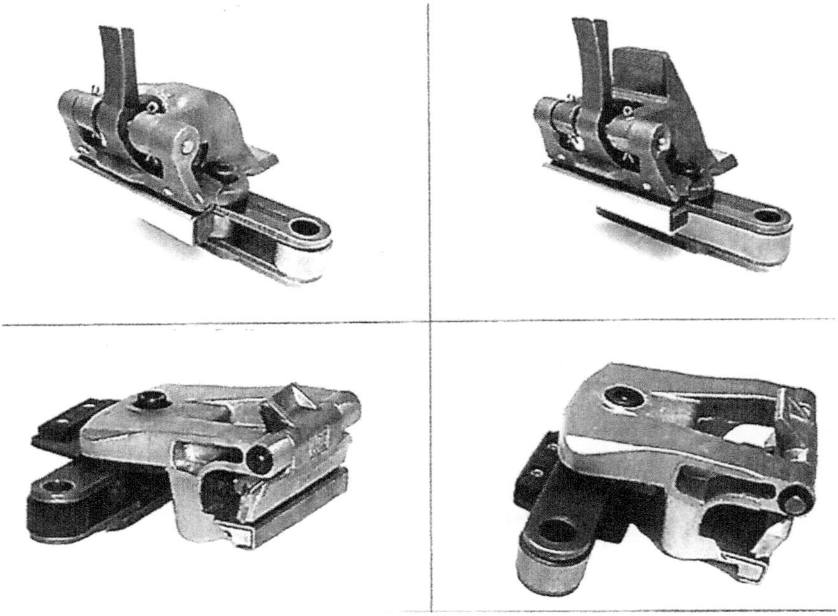

**Figure 1.23** Different types of tenter frame clips. Courtesy STM, Inc.

controlled by varying the divergence of the edge-clip paths. At the end of the frame the clips let go of the film, which is now three to eight times wider. The clips then wrap around a wheel and return to grab the film again at the entrance. The stretched film is wound up on a roll.

### 1.3.5.3 Biaxial Orientation

BO is stretching of the film in both machine and transverse directions, which is carried out in a two-step process [39]. First, it is stretched in the machine direction by passing it around heated rolls rotating at controlled and increasing speeds as described in Section 1.3.5.1. After machine direction stretching is completed, the film enters the tenter frame.

The overall schematic for the production of biaxially oriented film is in Fig. 1.24 [37]. This figure also shows the typical width profile of the film in this *sequential* orientation process. Mechanical properties of biaxially oriented films depend mainly on the amount of stretching and the ratio of the amounts of individual orientations.

It should be noted that the alternative to this sequential BO, which is widely used, there are known *simultaneous* methods, which involve complex movements of the film-edge grips so that the film is stretched in the

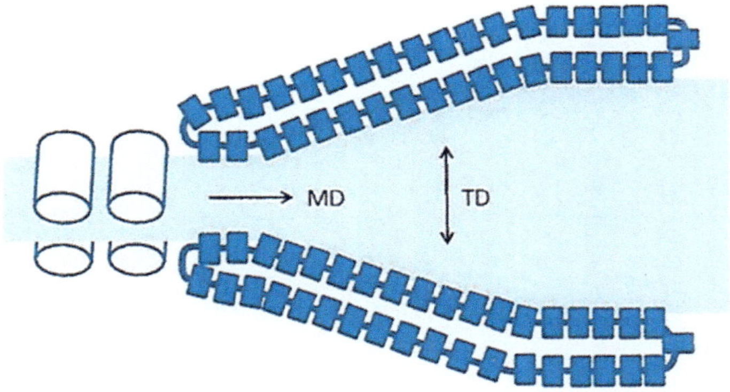


**Figure 1.24** Schematic of the overall biaxial orientation; MD: stretching between nipped rolls of differing speeds; TD: stretching with clamps [37].

**Figure 1.25** Industrial biaxial orientation system. Courtesy Marshall and Williams Plastics.

machine and transverse (cross-machine direction) at the same time. However, the process is mechanically complicated, and it is difficult to adjust the balance between the stretch directions [38], and therefore as it stands now, the sequential process is currently preferred for most films. A modern industrial biaxial orientation system is shown in Fig. 1.25.

#### *1.3.5.4 Blown Film Orientation*

Blown film may be biaxially orientated using the process called "double bubble" (see Fig. 1.26). The film is first formed in the die, and

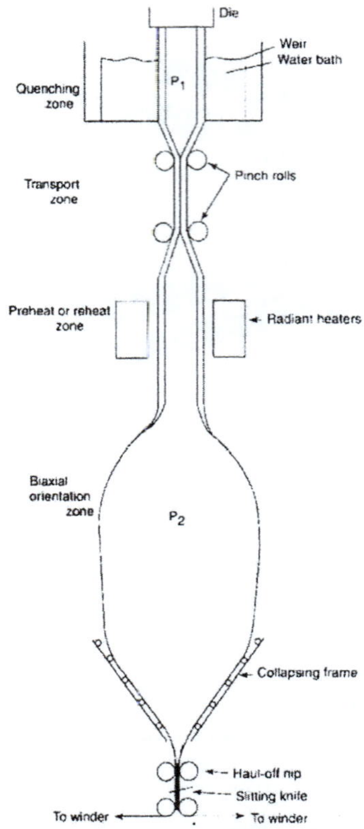


**Figure 1.26** Schematic of biaxial orientation during blown film process (double bubble) [40].

then it is inflated by air into a water quench. The film is reheated by radiant heater and then passed into the second bubble, which stretches the film in both directions [40]. This process is commonly used to make heat shrink films and bags. Second approach to orientation of blown film is by using counterrotating die described in Ref. [40].

The described machine orientation processes are relatively compact and are often put in-line with film producing lines.

# References

[1] Specialty Films, Industry Study #3732, Freedonia Group, 2019. <www. freedonia.com>.

[2] J.G. Drobny, Technology of Fluoropolymers, second ed., CRC Press, Boca Raton, FL, 2009, p. 1.

[3] R.J. Plunkett, US Patent 2,230,654, to Kinetic Chemicals, Inc., 1941.

[4] T.L. Miller, Personal Communication, 1998.
[5] J.G. Drobny, Technology of Fluoropolymers, second ed., CRC Press, Boca Raton, FL, 2009, p. 2.
[6] C. Maier, T. Calafut, Polypropylene, The Definitive User's Guide and Databook, Chapter 16, PDL Handbook Series, William Andrew Inc, Norwich, NY, 2000.
[7] S. Ebnesajjad, Fluoroplastics, Volume 2, Melt Processable Fluoropolymers, The Definitive User's Guide and Databook, Chapter 8, PDL Handbook Series, William Andrew Inc, Norwich, NY, 2003.
[8] W. Michaeli, (Chapter 5) Extrusion Dies for Plastics and Rubber, second ed., Hanser Publishers, Munich, 1992.
[9] M.L. Berins, Chapter 4 in: M.L. Berins (Ed.), Engineering Handbook of the SPI, fifth ed., Van Nostrand Reinhold, New York, 1991.
[10] F. Johannaber, Injection Molding Machines, A User's Guide, third ed., Hanser Publishers, Munich, 1994.
[11] S. Ebnesajjad, Surface Treatment of Materials for Adhesion Bonding, William Andrew, Inc., Norwich, NY, 2006, p. 97.
[12] S. Ebnesajjad, Surface Treatment of Materials for Adhesion Bonding, William Andrew, Inc., Norwich, NY, 2006, p. 110.
[13] S. Ebnesajjad, Surface Treatment of Materials for Adhesion Bonding, William Andrew, Inc., Norwich, NY, 2006, p. 111.
[14] S. Ebnesajjad, Surface Treatment of Materials for Adhesion Bonding, William Andrew, Inc., Norwich, NY, 2006, p. 114.
[15] S. Ebnesajjad, Surface Treatment of Materials for Adhesion Bonding, William Andrew, Inc., Norwich, NY, 2006, p. 115.
[16] S. Ebnesajjad, Surface Treatment of Materials for Adhesion Bonding, William Andrew, Inc., Norwich, NY, 2006, p. 121.
[17] D. Briggs, C. Kendall, Int. J. Adhes. Adhes. 2 (1982) 13.
[18] S. Ebnesajjad, Chapter 8 Polyvinyl Fluoride—Technology and Applications of PVF, Elsevier, Oxford, UK, 2013p. 200.
[19] S. Ebnesajjad, Polyvinyl Fluoride—Technology and Applications of PVF, Elsevier, Oxford, UK, 2013, p. 202.
[20] S. Ebnesajjad, Chapter 15 Fluoroplastics, Volume 1, Non-Melt Processible Fluoropolymers, The Definitive User's Guide and Data Book, second ed., Elsevier, Oxford, UK, 2015.
[21] Fluortek. <www.fluortek.com>; Acton Technologies. <www.acton.com>.
[22] Fluortek. www.fluortek.com; Technetic Group. <https://techneticsptfe.com/etching>.
[23] L.W. McKeen, Film Properties of Plastics and Elastomers, fourth ed., Elsevier, Oxford, UK, 2017, p. 75.
[24] L.W. McKeen, Film Properties of Plastics and Elastomers, fourth ed., Elsevier, Oxford, UK, 2017, p. 76.
[25] Lenderink Technologies. <www.lenderink.com/laminations>, 2019.

[26] R. Halle, Proceedings of TAPPI Conference, 1989, pp. 799–806.
[27] F.C. Stehling, P. Meka, J. Appl. Polym. Sci. 51 (1994) 105–119.
[28] J.M. Farley, P. Meka, J. Appl. Polym. Sci. 51 (1994) 121–131.
[29] The Chemours Company, Techniques for fabricating Teflon™ FEP film, in: Document C-10235 (3/16), 2016. <www.chemours.com>.
[30] P.A. Harper, E.M. Petrie, Plastics Materials and Processes—A Concise Encyclopedia, Van Nostrand Reinhold Co, New York, 1986, p. 344.
[31] P.A. Harper, E.M. Petrie, Plastics Materials and Processes—A Concise Encyclopedia, Van Nostrand Reinhold Co, New York, 1986, p. 345.
[32] P.A. Harper, E.M. Petrie, Plastics Materials and Processes—A Concise Encyclopedia, Van Nostrand Reinhold Co, New York, 1986, p. 346.
[33] L.W. McKeen, Film Properties of Plastics and Elastomers, fourth ed., Elsevier, Oxford, UK, 2017, p. 481.
[34] Machine Design. <www.machinedesign.com>, 2002.
[35] Dunmore Corporation. <www.dunmore.com>, 2019.
[36] Electrocube. <www.electrocube.com>, 2019.
[37] B.A. Morris, The Science and Technology of Flexible Packaging, Elsevier, Oxford, UK, 2017, p. 45.
[38] S. Ebnesajjad, Polyvinyl Fluoride, Technology and Applications, Elsevier, Oxford, UK, 2013, p. 123.
[39] L.W. McKeen, Film Properties of Plastics and Elastomers, fourth ed., Elsevier, Oxford, UK, 2017, p. 77.
[40] L.W. McKeen, Film Properties of Plastics and Elastomers, fourth ed., Elsevier, Oxford, UK, 2017, p. 78.

# 2 Materials for Fluoropolymer Films and Sheets

## 2.1 Perfluoroethylene (Polytetrafluoroethylene)

In addition to the presence of stable C–F bonds the polytetrafluoroethylene (PTFE) molecule possesses other features, which leads to materials of outstanding heat resistance, chemical resistance, and electrical insulation characteristics and with a low coefficient of friction. It is today produced by a number of mainly global chemical manufacturers.

### 2.1.1 Industrial Process for Production of Polytetrafluoroethylene

PTFE is produced by polymerization of tetrafluoroethylene (TFE) via the free radical addition mechanism in aqueous medium with water-soluble free radical initiators, such as peroxydisulfates, organic peroxides, or reduction–activation systems [1]. The additives have to be selected very carefully since they may interfere with the polymerization. They may either inhibit the process or cause chain transfer that leads to inferior products. When producing aqueous dispersions, highly halogenated emulsifiers, such as modified fluorinated acids [2] are used. TFE polymerizes readily at moderate temperatures (40°C–80°C or 104°F–176°F) and moderate pressures (0.7–2.8 MPa) (102–406 psi). The reaction is extremely exothermic (the heat of polymerization is 41 kcal/mol). In the absence of air, it disproportionates violently to yield carbon and carbon tetrafluoride. This decomposition generates the same energy as an explosion of black powder. The decomposition is initiated thermally; therefore the equipment used in handling and polymerization must be without hot spots.

In principle, there are two distinct methods of polymerization of TFE. When little or no dispersing agent is used and the reaction mixture is agitated vigorously, a precipitated granular polymer is produced. If proper type and sufficient amount of dispersant is used and mild agitation is maintained, the resulting product consists of small negatively charged oval-shaped colloidal particles (longer dimension less than

**Applications of Fluoropolymer Films.** DOI: https://doi.org/10.1016/B978-0-12-816128-9.00002-7

0.5 μm). The two products are distinctly different, even though both are high-molecular-weight PTFE polymers. The resulting aqueous dispersion can be either used for the production of fine PTFE powders or further concentrated into products used for direct dipping; coating, etc.

*Granular PTFE resins* are produced by polymerizing TFE alone or with a trace of comonomers with initiator and sometimes in the presence of an alkaline buffer in aqueous suspension medium. The product from the autoclave can consist of a mixture of water with particles of polymer of variable size and irregular shape. After the water is removed from the mixture the polymer is dried, and to obtain usable product, it is disintegrated before or after drying. Finished products usually have mean particle size 10–700 μm and apparent density 200–700 g/L, depending on grade [3].

*Fine-powder PTFE resins* are prepared in the following fashion: The first step in their manufacture is to prepare an aqueous colloidal *dispersion* by polymerization with initiator and emulsifier present [4]. Although the polymerization mechanism is not a typical emulsion type, some of the principles of emulsion polymerization apply here. Both the process and the ingredients have significant effects on the product. The solid contents of such dispersions can be as high as 40% by weight (approximately 20% by volume, because of the high density of PTFE). The dispersion has to be sufficiently stable through the polymerization not to coagulate in the autoclave yet unstable enough to allow subsequent controlled coagulation into fine powders. Gentle stirring ensures the stability of the dispersion. The finished dispersion is then diluted to a solids content of about 10% by weight and coagulated by controlled stirring and the addition of an electrolyte. The thin dispersion first thickens to give a gelatinous mass; then the viscosity decreases again, and the coagulum changes to air-containing, water-repellent agglomerates that float on the aqueous medium [5]. The agglomerate is dried gently; shearing must be avoided. The finished powder consists of agglomerates of colloidal particles with the mean size of 300–700 μm and has an apparent density in the range between 350 and 600 g/L. Transportation and handling of PTFE fine powders should be done below 19°C (66.2°F), the transition temperature to prevent particle fibrillation [6].

*PTFE aqueous dispersions* are made by the polymerization process used to make fine powders. Raw dispersions are polymerized to different particle sizes [7]. The optimum particle size for most applications is about 0.2 μm. The dispersion from the autoclave is stabilized by the addition of nonionic or anionic surfactants, followed by concentration to a solid content of 60%–65% by electrodecantation, evaporation, or thermal concentration [8]. After further modification with chemical additives the commercial product is sold with a polymer content of about 60% by weight, viscosity of several centipoises, and specific gravity around 1.5. The processing characteristics of the dispersion depend on the conditions for the polymerization and the type and amounts of the chemical additives contained in it. Stability is a key requirement of the final product. Typically, a PTFE aqueous dispersion must have a shelf life several months to 1 year. It should withstand transportation and handling during processing. The shear rate during processing must be low enough as to not causing the agglomeration of the particles. Ideally, the temperature during transportation, storage, and processing should be below 19°C (66.2°F), for the same reason as it is in case of fine powders, namely to prevent particle fibrillation. Processing of aqueous PTFE dispersions is discussed in more detail in Sections 3.1.3 and 3.1.4.

## 2.1.2 Structure and Related Properties of Polytetrafluoroethylene

PTFE is a linear polymer free from any significant amount of branching (Fig. 2.1). While the molecule of polyethylene is in the form of planar zigzag in the crystalline zone, this is sterically impossible with that

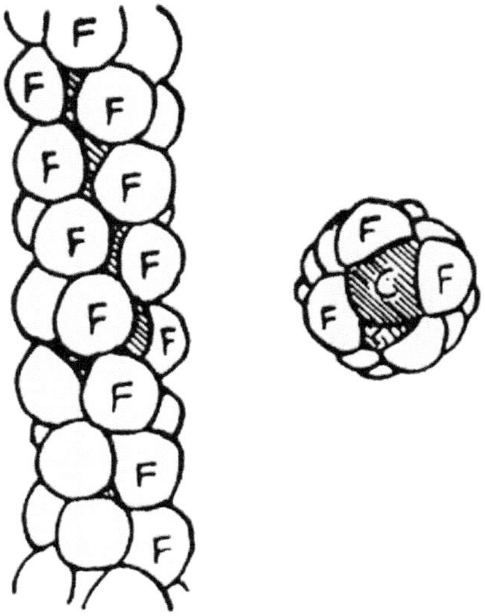


**Figure 2.1** Schematic representation of the PTFE helix. Reprinted from G.P. Koo, in: L.A. Wall (Ed.), Fluoropolymers, Wiley-Interscience, New York, 1972, p. 508. With permission.

of PTFE due to the fluorine atoms being larger than those of hydrogen. As a consequence, the molecule takes up a twisted zigzag, with the fluorine atoms packing tightly in a spiral around the carbon–carbon skeleton [9]. A complete turn of the spiral will involve over 26 carbon atoms below 19°C (66.2°F) and 30 above it, there being a transition point involving a 1% volume change at this temperature. The compact interlocking of the fluorine atoms leads to a molecule of great stiffness, and it is this feature that leads to the high crystalline melting point and thermal form stability of the polymer. The outer sheath of fluorine atoms protects the carbon backbone, thus providing the chemical inertness and stability. It also lowers the surface energy, giving PTFE a low coefficient of friction and nonstick properties [10].

The intermolecular attraction between PTFE molecules is very small, the computed solubility parameter being 12.6 $(MJ/m^3)^{1/2}$. The polymer in bulk does not thus have the high rigidity and tensile strength, which is often associated with polymers with a high softening point.

The carbon–fluorine bond is very stable. Further, where two fluorine atoms are attached to a single carbon atom, there is a reduction in the C–F bond distance from 1.42 to 1.35 Å. As a result, bond strengths may be as high as 504 kJ/mol. Since the only other bond present is the stable C–C bond, PTFE has very high heat stability, even when heated above its crystalline melting point of 327°C (620.6°F). Because of its high crystallinity and incapability of specific interaction, there are no solvents at room temperature. At temperatures approaching the melting point, certain fluorinated liquids, such as perfluorinated kerosenes, will dissolve the polymer.

The properties of PTFE are dependent on the type of polymer and the method of processing. The polymer may differ in particle size and/or molecular weight. The particle size will influence ease of processing and the quantity of voids in the finished product while the molecular weight will influence crystallinity and hence many physical properties. The processing techniques will also affect both crystallinity and void content.

The weight average molecular weights of commercial PTFE polymers are in the range $1-5 \times 10^6$ [11], and the percentage of crystallinity is greater than 94% as manufactured. Fabricated parts are less crystalline. The degree of crystallinity of the finished product will depend on the rate of cooling from the processing temperatures. Slow cooling will lead to high crystallinity, with fast cooling giving the opposite effect. Low-molecular-weight materials will also be more crystalline. Fig. 2.2 shows the relationship between percentage crystallinity and specific gravity at 23°C (73.4°F).

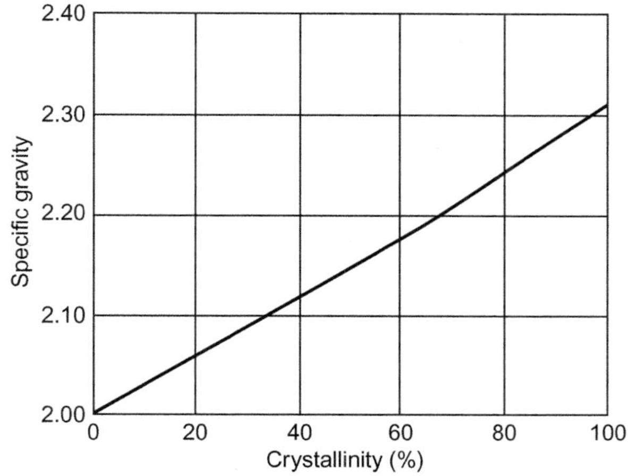


**Figure 2.2** Specific gravity as a function of crystallinity level. Courtesy ICI Ltd.

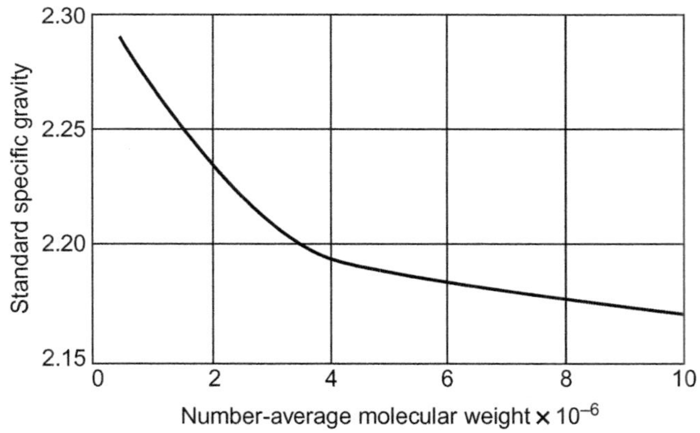


**Figure 2.3** Standard specific gravity as a function of molecular weight. Courtesy ICI Ltd.

It is observed that the dispersion polymer, which is of finer particle size and lower molecular weight, gives products with a vastly improved resistance to flexing and also distinctly higher tensile strengths. These improvements appear to arise through the formation of fiber-like structures in the mass of polymer during processing. The effect of molecular weight on the standard specific gravity is shown in Fig. 2.3.

### 2.1.3 General Properties of Polytetrafluoroethylene

PTFE is a tough, flexible, nonresilient material of moderate tensile strength but with excellent resistance to heat, chemicals, and to the passage of an electric current. It remains ductile in compression at temperatures as low as 4K (−269°C or −452°F). As with other plastics materials, temperature has a considerable effect on mechanical properties. This is clearly illustrated in Fig. 2.4 in the case of stress to break and elongation at break. Even at 20°C (68°F), unfilled PTFE has a measurable creep with compression loads as low as 300 psi (2.1 MPa).

The coefficient of friction is unusually low and stated to be lower than that of any other solid. A number of different values have been quoted in the literature but are usually in the range 0.02–0.10 for polymer to polymer.

PTFE is an outstanding insulator over a wide range of temperature and frequency. The volume resistivity (100 seconds value) exceeds $10^{20}$ Ωm, and it appears that any current measured is a polarization current rather than conduction current. The power factor is negligible in the temperature range −60°C to +250°C (−76°F to 482°F) at frequencies up to $10^{10}$ Hz. The polymer has a low dielectric constant similarly unaffected by frequency. The only effect of temperature is to alter the density which has been found to influence the dielectric constant according to the relationship

$$\text{Dielectric constant} = \frac{1 + 0.238D}{1 - 0.119D}$$

where $D$ is the specific gravity.

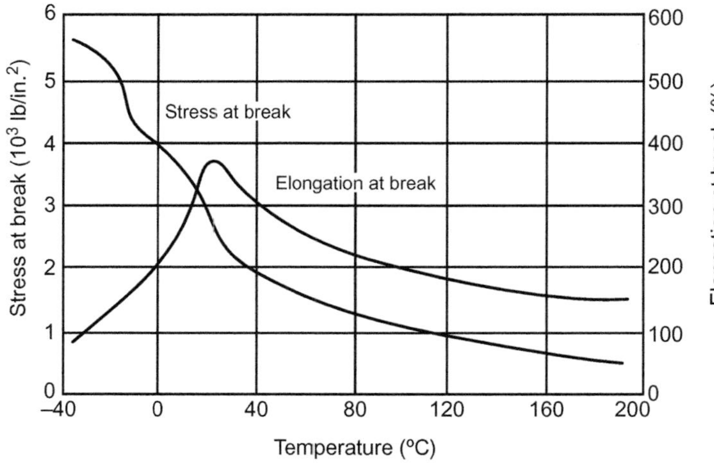


**Figure 2.4** Effect of temperature on stress at break and elongation at break. Courtesy ICI Ltd.

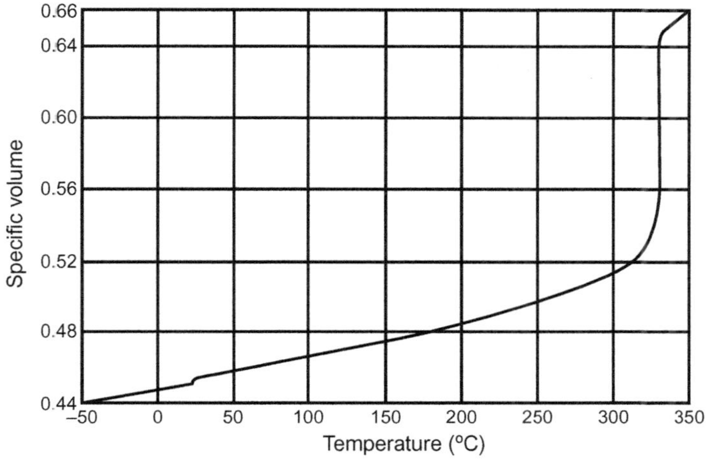


**Figure 2.5** Variation of specific volume of PTFE with temperature. Courtesy ICI Ltd.

Fig. 2.5 shows the influence of temperature on specific volume (reciprocal specific gravity). The exact form of the curve is somewhat dependent on the crystallinity and the rate of temperature change. A small transition is observed at about 19°C (66.2°F) and a first-order transition (melting) at about 327°C (620.6°F). Above this temperature the material does not exhibit true flow but is rubbery. A melt viscosity of $10^{10}-10^{12}$ P ($10^{9}-10^{11}$ Pa s) [10] has been measured at about 350°C (662°F). A slow rate of decomposition may be detected at the melting point, and this increases with a further increase in temperature. Processing temperatures, except possibly in the case of extrusion, are, however, rarely above 380°C (716°F).

The chemical resistance of PTFE is exceptional. There are no solvents, and it is attacked at room temperature only by molten alkali metals and in some cases by fluorine. Treatment with a solution of sodium metal in liquid ammonia will sufficiently alter the surface of a PTFE sample to enable it to be cemented to other materials using epoxide resin adhesives.

Although it has good weathering resistance, PTFE is degraded by ionizing radiation (gamma rays, X-rays, electron beam radiation). The degradation of the polymer in the air and oxygen due to chain scission and is fairly rapid. Such scission results in molecular weight reduction. When irradiated by electron beam, the molecular weight is reduced up to six orders of magnitude [12].

The polymer is not wetted by water and does not measurably absorb it. The permeability to gases is low, the water vapor transmission rate being approximately half of that of low-density polyethylene and poly (ethylene terephthalate) (PETG). Voids greater than molecular size increase the permeation [13].

### 2.1.4 Forms of PTFE Resins for Films and Sheets

Commercial PTFE resins mentioned in Section 2.1.2 used normally for PTFE films are available in the following three forms:

1. Granular resins with median particle size of 300 and 600 μm.
2. Fine powders (dispersion polymer) obtained by coagulation of a dispersion. It consists of agglomerates with an average diameter of 450 μm made up of primary particles 0.1 μm in diameter.
3. Dispersions (lattices) containing about 60% polymer in particles with an average diameter of about 0.16 μm.

Each of these forms has a specific way to be processed. The corresponding processing methods used for the manufacture of PTFE films and sheets are discussed in detail in Chapter 3, Polytetrafluoroethylene Films.

### 2.1.5 Modified Polytetrafluoroethylene

PTFE has many remarkable properties (see Sections 2.1.3 and 2.1.4), but it has several shortcomings, which limit its utility as an engineering material. It exhibits a significant cold flow (low creep resistance), is difficult to weld, and contains a large number of microvoids due to a rather poor coalescence of particles during the sintering process. The weaknesses result from the combination of a high molecular weight (extremely high melt viscosity) and a high degree of crystallinity. To reduce these disadvantages, modified PTFE, which contains a small amount (0.01–0.1 mol.%) of a comonomer was developed. The most suitable comonomer was found to be perfluoropropylvinyl ether (PPVE) [14]. The comonomer reduces the degree of crystallinity and the size of lamellae. The polymerization process is similar to that for standard PTFE except additives to control the molecular weight are used.

The resulting polymer has a melt viscosity lower by an order of magnitude and because of that the particles coalesce better during sintering.

Moreover, it has a markedly improved weldability [15]. All the important physical properties of commercial modified PTFE are significantly improved without any noticeable reduction of other properties. Modified PTFE is still processed by the same techniques as the standard polymer and no adjustments to processing techniques are necessary. Because of the lower melt viscosity, modified PTFE performs better in coined molding, blow molding, and thermoforming. Currently available grades of modified PTFE are granular molding resins, fine powders, and aqueous dispersions. Properties of modified PTFE and the conventional PTFE are compared in Table 2.1.

### 2.1.6 Processing of PTFE

Because PTFE has unique properties and melting behavior (see Section 2.1.3), it is processed by different techniques than most thermoplastics. The main difference is due to the very high melt viscosity (typically$10^9-10^{10}$ Pa s) [10], which prevents any significant melt flow. As pointed out earlier, PTFE is manufactured and offered to the market in essentially three forms, namely as granular resins, as fine powders, and as aqueous dispersions. Although they are all chemically high-molecular-weight PTFE with an extremely high melt viscosity (see earlier), each of them requires a different processing technique. *Granular resins* are processed by compression molding, isostatic molding, and ram extrusion. Most of the molded and extruded resins are sintered at temperatures above their crystalline melting temperature [10]. Some proportion of molded and sintered resins is used for the production of skived films (see Section 3.1.1).

*Fine powders* are processed mainly by extrusion and calandering. Finished products from fine powders may be either unsintered or sintered (see Section 3.1.2).

*Aqueous dispersions* (containing typically 60% solids) are mostly diluted to required solids content for the given process and/or product and used for cast unsupported and supported films as well as for coated woven or nonwoven fabrics and cast films [16] (see Sections 3.1.3 and 3.1.4). The majority of coated fabrics and cast films are sintered to assure optimum mechanical properties.

### 2.1.7 Applications for Polytetrafluoroethylene

About one half of the PTFE resin produced is used in electrical and electronic applications [17] with major use for insulation of hookup

**Table 2.1 Comparison of Modified PTFE and Conventional PTFE.**

| Property | ASTM Test Method | Modified PTFE | Conventional PTFE |
|---|---|---|---|
| Tensile strength (MPa) | D4894 | 31 | 34 |
| Elongation at break (%) | D4894 | 450 | 375 |
| Specific gravity | D4894 | 2.17 | 2.16 |
| Deformation under load at 23°C (%) | D695 | | |
| 3.4 MPa | | 0.2 | 0.7 |
| 6.9 MPa | | 0.4 | 1.0 |
| 13.8 MPa | | 3.2 | 8.2 |
| Deformation under load (%) | D695 | | |
| 6.9 MPa at 23°C | | 5.3 | 6.7 |
| 3.4 MPa at 100°C | | 5.4 | 8.5 |
| 1.4 MPa at 200°C | | 3.6 | 6.4 |
| Void content of typical parts (%) | FTIR | 0.5 | 1.5 |
| Dielectric strength, kV/mm (76 μm film) | D149 | 208 | 140 |
| Weld strength[a] (%) | D4894 | 66–87 | Very low |
| Permeation of perchloroethylene | Comparative rates | | |
| Vapor | | 2 | 5 |
| Liquid | | 4 | 13 |
| Permeation of hexane | Comparative rates | | |
| Vapor | | 0.2 | 3.4 |
| Liquid | | 0 | 23.4 |

[a]Specimens welded after sintering.
*Source*: Modified from J.G. Drobny, in: M. Gilbert (Ed.), Brydson's Plastics Materials, eighth ed., Elsevier, Oxford, 2017, p. 391.

wire for military and aerospace electronic equipment. PTFE is also used as insulation for airframe and computer wires, as "spaghetti" tubing, and in electronic components.

Large quantities of PTFE are used in the chemical industry in fluid-conveying systems as gaskets, seals, molded packing, bellows, hose, and lined pipe [17] and as lining of large tanks or process vessels. PTFE is also used in laboratory apparatus.

Unsintered tape is used for sealing threads of pipes for water and other liquids and for wrapping coaxial cables. Pressure-sensitive tapes with silicone or acrylic adhesives are made from skived or cast PTFE films.

Because of its very low friction coefficient, PTFE is used for bearings, ball-, and roller-bearing components, and sliding bearing pads in static and dynamic load supports [17]. Piston rings of filled PTFE in nonlubricated compressors permit operation at lower power consumption or at increased capacities [18].

Modified PTFE (e.g., Teflon NXT), because of its improved processing, lower creep, improved permeation, less porosity and better insulation than standard PTFE, finds use in pipe and vessel linings, gaskets and seals, fluid-handling components, wafer processing, and electric and electronic industries. Since PTFE is highly inert and nontoxic, it finds use in medical applications for cardiovascular grafts, heart patches, ligaments for knees, and others [18].

Highly porous membranes are prepared by a process based on the fibrillation of high-molecular-weight PTFE [19] (see Section 3.1.5). Since they have a high permeability for water vapor and none for liquid water, it is combined with fabrics and used for breathable waterproof garments and camping gear (made by W.L. & Associates and sold under the brand name GORE-TEX). Other uses for these membranes are for special filters, analytical instruments, and in fuel cells [20].

Because of its low surface energy and limited chemical reactivity, PTFE exhibits poor wettability and adhesive bonding. Surface modification of PTFE is an established and valuable technique to adjust the surface properties to improve wetting and adhesive bonding. Although many different methods for that have been developed, such as radiation grafting [21], plasma treatment [22], roughness [23], impregnation with metal oxides [24], the well-established commercial method is the treatment with sodium–naphthalene complex [25,26].

## 2.2 Melt-Processible Fluoropolymers

Commercial thermoplastic fluoropolymers with the exception of perfluoroethylene (PTFE) and polyvinyl fluoride (PVF) are processed as melts into films, sheets, profiles, and moldings using conventional manufacturing methods. They are widely used in chemical, automotive, electrical, and electronic industries; in aircraft and aerospace; in communications, construction, medical devices, special packaging, protective garments, and a variety of other industrial and consumer products.

**Table 2.2 Categories of Commercial Fluoropolymers.**

| Type | Form | Partially Fluorinated | Perfluorinated |
|---|---|---|---|
| Crystalline | Resin | ETFE | PTFE |
| | | PVDF | FEP |
| | | PVF | PFA, MFA |
| | | PCTFE | |
| | | ECTFE | |
| Amorphous | Resin | FEVE | Teflon AF |
| | | | CYTOP |
| | Elastomer | FKM | FFKM |

*CYTOP*, Cyclopolymer of perfluorodiene; *ECTFE*, copolymer of ethylene and chlorotrifluoroethylene; *ETFE*, copolymer of ethylene and tetrafluoroethylene; *FEP*, copolymer of tetrafluoroethylene and hexafluoropropylene; *FEVE*, fluorinated ethylene vinyl ether; *FFKM*, perfluorinated fluorocarbon elastomer; *FKM*, fluorocarbon elastomer; *MFA*, copolymer of tetrafluoroethylene and perfluoromethylvinyl ether; *PCTFE*, poly(chlorotrifluoroethylene); *PFA*, copolymer of tetrafluoroethylene and perfluoropropylvinyl ether; *PTFE*, polytetrafluoroethylene; *PVDF*, polyvinylidene fluoride; *PVF*, polyvinyl fluoride; *Teflon AF*, copolymer of tetrafluoroethylene and perfluoro-2,2dimethyl-1,3 dioxide.

Categories of commercial fluoropolymers are shown in Table 2.2 and a summary of properties of selected melt-processible fluoropolymers is in Table 2.3. These are homopolymers, copolymers, and terpolymers. Currently two of them are perfluorinated polymers, and the remainder are partially fluorinated polymers. The next two sections will cover the two melt-processible perfluoropolymers followed by the remaining melt-processible fluoropolymers.

## 2.2.1 Industrial Process for the Production of Melt-Processible Fluoropolymers

### *2.2.1.1 Industrial Process for the Production of Perfluoroalkoxy Resins*

There are several methods of copolymerization of hexafluoropropylene (HFP) and TFE using different catalysts at different temperatures [27–29] to produce fluorinated ethylene propylene (FEP). Aqueous and nonaqueous dispersion polymerizations appear to be the most convenient commercial routes. The conditions for this type of process are similar to those for the dispersion homopolymerization of TFE. FEP is

**Table 2.3 Summary of Properties of Selected Melt-Processible Fluoropolymers.**

| Property | Material | | | | |
|---|---|---|---|---|---|
| | PFA | FEP | PCTFE | ETFE | ECTFE |
| Specific gravity | 2.16 | 2.15 | 2.13 | 1.70 | 1.68 |
| Tensile strength (psi) | 4500 | 3000 | 4000 | 6500 | 7000 |
| (MPa) | 31 | 21 | 28 | 45 | 48 |
| Elongation at break (%) | 300 | 290 | 140 | 150 | 200 |
| Hardness, Shore D, or Rockwell | D60 | D55 | R109 | D75 | D75 |
| Melting temperature (°F) | 590 | 500–535 | 394 | 520 | 465 |
| (°C) | 310 | 260–280 | 201 | 271 | 240 |
| Max. continuous operation temperature (°F) | 500 | 400 | 350 | 350 | 340 |
| (°C) | 260 | 204 | 177 | 177 | 171 |
| Embrittlement temperature (°F) | NA | −100 | −423 | −150 | −105 |
| (°C) | – | −73 | −252 | −201 | −76 |
| Deflection temperature at 66 psi (°F) | NA | 158 | 258 | 220 | 240 |
| (°C) | – | 70 | 126 | 104 | 116 |
| Deflection temperature at 264 psi (°F) | NA | N | N | 160 | 170 |
| (°C) | – | – | – | 71 | 77 |
| Dielectric strength (kV/mil) | 4.0 | 6.5 | 3.5 | 7.0 | 2.0 |
| (kV/mm) | 160 | 260 | 140 | 280 | 80 |
| Dielectric constant at 1 kHz | 2.1 | 2.1 | 2.5 | 2.6 | 2.6 |
| Dissipation factor at 1 kHz | <0.0002 | <0.0002 | <0.025 | <0.0008 | <0.0015 |
| Water vapor permeability[a] | VH | H | L | M | VL |
| Chemical resistance[a] | VH | VH | M | M | M |
| Coefficient of friction[a] | L | M | H | H | H |

*ECTFE*, Copolymer of ethylene and chlorotrifluoroethylene; *ETFE*, copolymer of ethylene and tetrafluoroethylene; *FEP*, copolymer of tetrafluoroethylene and hexafluoropropylene; *PCTFE*, poly(chlorotrifluoroethylene).
[a]*Note*: *VL*, very low; *L*, low; *M*, moderate; *H*, high; *VH*, very high.

a random copolymer, that is, HFP units add to the growing chain at random intervals. The optimal composition of the copolymer is such that the mechanical properties are retained in the usable range and that it has low enough melt viscosity for an easy melt-processing [30]. Commercial FEP is available as low-melt-viscosity grades for injection molding, medium-viscosity and high-viscosity grades for extrusion, and aqueous dispersions with 55% of solids by weight [30,31].

#### 2.2.1.2 Industrial Process for the Production of Perfluoroalkoxy Resins

Perfluoroalkoxy resins are prepared by copolymerization of TFE and perfluoroalkoxy monomers in either aqueous or nonaqueous media [32–34].

#### 2.2.1.3 Industrial Process for the Production of ETFE Resins

Commercial products based on copolymers of ethylene and TFE are made by free radical–initiated addition copolymerization [35]. Small amounts (1–10 mol.%) of modifying comonomers are added to eliminate a rapid embrittlement of the product at exposure to elevated temperatures. Examples of the modifying comonomers are perfluorobutylethylene, HFP, perfluorovinyl ether, and hexafluoroisobutylene (HFIB) [36]. Additional information on the methods to prepare ETFE copolymers are in Ref. [37].

#### 2.2.1.4 Industrial Process for the Production of Poly (Chlorotrifluoroethylene) Resins

Polymerization may be carried out by techniques akin to those used in the manufacture of PTFE. More specifically, CTFE is polymerized by bulk, suspension, and emulsion techniques [38]. The tendency of poly(chlorotrifluoroethylene) (PCTFE) to become brittle during use can be reduced by incorporating small amounts (less than 5%) of vinylidene fluoride (VDF) during the polymerization process [39].

#### 2.2.1.5 Industrial Process for the Production of ECTFE Resins

The copolymerization of ethylene and CTFE (ECTFE) is performed as a free radical suspension process in aqueous media at low temperatures. Lowering the temperature reduces the number of ethylene blocks in the polymer backbone that are susceptible to thermal degradation. A commercial polymer with an overall CTFE-to-ethylene ratio of 1:1 contains ethylene blocks and CTFE blocks in the proportion lower than 10 mol.% each [40]. Reaction pressure is adjusted to give the desired copolymer ratio [41]. Typical pressures during the process are on the order of 3.5 MPa (508 psi). In some cases, modifying monomers are added to reduce high-temperature stress cracking of the pure ECTFE copolymer. The modified products typically have a lower degree of crystallinity and lower melting points [40].

During the copolymerization the product precipitates as a fine powder, with particles typically less than 20 μm in major dimension. These particles eventually agglomerate into roughly spherical beads and the reactor product is a mixture of beads and powder. The product is then dewatered and dried. It is further processed into extruded pellets for melt-processing (extrusion, injection molding, blow molding, etc.) or ground and screened into powder coating grades [42]. Additional methods of copolymerization of ethylene and CTFE are discussed in Ref. [43].

#### 2.2.1.6 Industrial Process for the Production of Polyvinylidene Fluoride Resins

The most common methods of producing homopolymers and copolymers of VDF are emulsion and suspension polymerizations, although other methods are also used [44].

*Emulsion polymerization* requires the use of free radical initiators, fluorinated surfactants, and often chain transfer agents. The polymer isolated from the reaction vessel consists of agglomerated spherical particles ranging in diameter from 0.2 to 0.5 μm [45]. It is then dried and supplied as a free-flowing powder or as pellets, depending on the intended use. If very pure polyvinylidene fluoride (PVDF) is required, the polymer is rinsed before the final drying to eliminate any impurities such as residual initiator and surfactants [46]. *Aqueous suspension polymerization* requires the usual additives, such as free radical initiators, colloidal dispersants (not always), and chain transfer agents, to control molecular weight. After the process is completed the suspension contains spherical particles approximately 100 μm in diameter. Suspension polymers are available as free-flowing powder or in pellet form for extrusion or injection molding.

The powdered polymers from emulsion or suspension polymerizations intended to be used for solvent-based coatings are often milled into finer particle size with higher surface area for easier dissolution when used as coatings for metal and other substrates [47].

Small amounts of comonomers (typically less than 6%) are often added to improve specific performance characteristics in cases where homopolymer is deficient. A higher level of comonomer than that (e.g., HFP) would yield a product with elastomeric characteristics [47]. VDF also copolymerizes with other monomers, such as acrylic compounds. Barium and strontium salts have been added to PVDF to improve its thermal stability [48].

#### *2.2.1.7 Industrial Process for the Production of THV Terpolymers*

Chemically, THV terpolymers are a class of terpolymers of TFE, HFP, and VDF produced by emulsion polymerization [49]. The resulting dispersion is either processed into powders and pellets or concentrated with emulsifier and supplied in that form to the market. At this writing, the manufacturer is Dyneon LLC (a 3M Company) and offers are essentially seven commercial grades (six dry grades in pellet or agglomerate form and one aqueous dispersion) available that differ in the monomer ratios and consequently in melting points, specific gravity, chemical resistance, optical properties, and flexibility. They marketed under the trade name Dyneon THV Fluoroplastic.

### 2.2.2 Structure and Related Properties of Melt-Processible Fluoropolymers

#### *2.2.2.1 Structure and Related Properties of Fluorinated Ethylene Propylene Resins*

Perfluorinated ethylene propylene (FEP) is a random copolymer, that is, HFP units add to the growing chain at random intervals. Commercial FEP is available as low-melt-viscosity grades for injection molding, grades for extrusion, medium-viscosity grades, high-viscosity grades, and as aqueous dispersions with 55% solids by weight [30,31].

The commercial polymers are mechanically similar to PTFE but with a somewhat greater impact strength. They also have the same excellent electrical insulation properties and chemical inertness. Weathering tests in Florida showed no change in properties after 4 years. The material also shows exceptional nonadhesiveness. The coefficient of friction of the resin is low but somewhat higher than that of PTFE. FEP films transmit more ultraviolet (UV), visible, and infrared radiation than ordinary window glass. They are considerably more transparent to the infrared and UV spectra than glass. The refractive index of FEP films is in the range 1.341–1.347 [50]. Films up to 0.010 in. (0.25 mm) thick show good transparency [51].

The maximum service temperature is about 60°C lower than that of PTFE for use under equivalent conditions. Continuous service at 200°C (392°F) is possible for a number of applications [51]. The polymer melts at about 260°C (500°F) [52]. The melting point is the only first-order transition observed in FEP. Melting increases the volume by 8% [53]. FEP resins have a very low–energy surface and are therefore very difficult to wet. The wetting can be improved by a solution of sodium in liquid ammonia or naphtalenyl sodium in tetrahydrofurane [54] or corona discharge [55].

#### 2.2.2.2 *Structure and Related Properties of Perfluoroalkoxy Resins*

Because of the high bond strength between carbon, fluorine, and oxygen atoms, PFA and MFA exhibit nearly the same unique properties as PTFE at temperatures ranging from extremely low to extremely high. Since they can be relatively easily processed by conventional methods for thermoplastics into film and sheets without microporosity, they have distinct advantage over PTFE in certain applications, such as corrosion protection and antistick coatings [56]. These polymers are semicrystalline, and the degree of crystallinity depends on the fabrication conditions, particularly on the cooling rate.

Commercial grades of PFA melt typically in the temperature range from 300°C to 315°C (572°F to 599°F) depending on the content of PPVE. The degree of crystallinity is typically 60% [57].

There is only one first-order transition at −5°C (23°F) and two second-order transitions, one at 85°C (185°F) and the other at −90°C (−130°F) [57].

In general, mechanical properties of PFA are very similar to those of PTFE within the range from −200°C to +250°C (−328°F to +482°F). The mechanical properties of PFA and MFA at room temperature are practically identical; differences become obvious only at elevated temperatures, because of the lower melting point of MFA.

In contrast to PTFE with measurable void content the melt-processed PFA is intrinsically void free. As a result, lower permeation coefficients should result because permeation occurs by molecular diffusion. This is indeed the case, but the effect levels off at higher temperatures [58].

The most remarkable difference between PTFE and PFA is the considerably lower resistance to deformation under load (cold flow) of the latter. In fact, addition of even minute amounts of PFA to PTFE improves its resistance to cold flow [57].

PFA and MFA exhibit considerably better electrical properties than most traditional plastics. In comparison with the partially fluorinated polymers, they are only slightly affected by temperature up to their maximum service temperature [59]. The dielectric constant remains at 2.04 over a wide range of temperature and frequencies (from 100 Hz to 1 GHz). The dissipation factor at low frequencies (from 10 Hz to 10 kHz) decreases with increasing frequency and decreasing temperature. In the range from 10 kHz to 1 MHz, temperature and frequency have little effect; while above 1 MHz the dissipation factor increases with the frequency [60].

Generally, fluorocarbon films exhibit high transmittance in the UV, visible, and infrared regions of the spectrum. This property depends on the degree of crystallinity and the crystal morphology in the polymer. For example, 0.025-mm (0.001-in.) thick PFA film transmits more than 90% of visible light (wavelength 400–700 nm). A 0.2-mm (0.008-in.) thick MFA film was found to have a high transmittance in the UV region (wavelength 200–400 nm). The refractive indexes of these films are close to 1.3 [61].

PFA and MFA have an outstanding chemical resistance even at elevated temperatures. They are resistant to strong mineral acids, inorganic bases, and inorganic oxidizing agents and to most of the organic compounds and their mixtures common in the chemical industry. However, they react with fluorine and molten alkali [60].

Elemental sodium, as well as other alkali metals, reacts with perfluorocarbon polymers by removing fluorine from them. This reaction has a practical application for improving surface wettability and adhesive bonding of perfluorocarbon polymers to other substrates [62].

The absorption of water and solvents by perfluoropolymers is in general very low [62]. Permeability is closely related to absorption and depends on temperature, pressure, and the degree of crystallinity. Since these resins are melt-processed, they are usually free of voids and, therefore, exhibit much lower permeability than PTFE. Permeation through PFA occurs via molecular diffusion [60].

#### 2.2.2.3 Structure and Related Properties of ETFE Resins

ETFE resins are essentially alternating copolymers [63], and in the molecular formula, they are isomeric with PVDF with a head-to-head, tail-to-tail structure. However, in many important physical properties, the modified ETFE copolymers are superior to PVDF with the exception of the latter's remarkable piezoelectric and pyroelectric characteristics. The continuous service temperature of ETFE is 150°C (302°F) [64]. It exhibits exceptional toughness and abrasion resistance, flex and creep resistance, and high tensile strength. The strength can be further increased by cross-linking by peroxide or electron beam irradiation [65]. Modified ETFE has excellent resistance to common solvents and chemicals [66]. As normally produced, ETFE has about 88% alternating sequences and a melting temperature of 270°C (518°F) [67].

#### *2.2.2.4 Structure and Related Properties of Poly (Chlorotrifluoroethylene) Resins*

The major differences in properties between PTFE and PCTFE can be related to chemical structure. The introduction of a chlorine atom, which is larger than the fluorine atom, breaks up the very neat symmetry, which is shown by PTFE and thus reduces the close chain packing. It is still, however, possible for the molecules to crystallize, albeit to a lower extent than PTFE [68]. The introduction of the chlorine atom in breaking up the molecular symmetry appears to increase the chain flexibility, and this leads to a lower softening point. On the other hand the higher interchain attraction results in a harder polymer with a higher tensile strength. The unbalanced electrical structure adversely affects the electrical insulation properties of the material and limits its use in high-frequency applications. Because of the lower tendency to crystallization, it is possible to produce thin transparent films [68].

The chemical resistance of PCTFE is good but not as good as that of PTFE. Under certain circumstances substances such as chlorosulfonic acid, molten caustic alkalis and molten alkali metal will adversely affect the material. Alcohols, acids, phenols, and aliphatic hydrocarbons have little effect but certain aromatic hydrocarbons, esters, halogenated hydrocarbons, and ethers may cause swelling at elevated temperatures.

The polymer melting at 211°C (411.8°F) [68] and above shows better cohesion of the melt than PTFE. It may be processed by conventional thermoplastics processing methods at temperatures in the range 230°C–290°C (446°F–554°F) [69].

The homopolymers and copolymers with VDF exhibit outstanding barrier properties [70]. PCTFE does not absorb visible light, and it is possible to produce optically clear sheets and parts up to 3.2 mm (1/8 in.) thick by quenching from melt [70,71]. PCTFE alone has a good resistance to ionizing radiation that is further improved by copolymerization with small amounts of VDF [71]. The disadvantage of PCTFE is that it is attacked by many organic materials and has a low thermal stability in the molten state [71].

#### 2.2.2.5 Structure and Related Properties of ECTFE Resins

Commercial polymer with an overall CTFE-to-ethylene ratio of 1:1 contains ethylene blocks and CTFE blocks of less than 10 mol.% each. The modified copolymers produced commercially also exhibit improved high-temperature stress cracking. Typically, the modified copolymers are less crystalline and have lower melting points [64]. Modifying monomers are HFIB, perfluorohexylethylene, and PPVE [72].

ECTFE resins are tough, moderately stiff, and creep resistant with service temperatures from −100°C to +150°C (−148°F to +302°F). The melt temperature depends on the monomer ratio in the polymer and is in the range of 235°C–245°C (455°F–473°F) [73]. Its chemical resistance is good and similar to PCTFE. ECTFE, as most fluoropolymers, has an outstanding weathering resistance. It also resists high-energy gamma and electron beam radiation up to 100 Mrad (1000 kGy) [73].

#### 2.2.2.6 Structure and Related Properties of Polyvinylidene Fluoride Resins

Commercial products based on PVDF contain various amounts of comonomers such as HFP, CTFE, and TFE that are added at the start of the polymerization to obtain products with different degrees of crystallinity. Products based on such copolymers exhibit higher flexibility, chemical resistance, elongation, solubility, impact resistance, optical clarity, and thermal stability during processing. However, they often have lower melting points, higher permeation, lower tensile strength, and higher creep than the PVDF homopolymer [74]. Barium and strontium salts have been added to PVDF to improve its thermal stability.

Some of the important properties of PVDF homopolymers and copolymers are a function of the crystalline content and type of crystalline structure, both of which are affected by the processing methods and conditions. PVDF exhibits a complex crystalline polymorphism, which cannot be found in other known synthetic polymers. There are a total of four distinct crystalline forms: alpha, beta, gamma, and delta. These are present in different proportions in the material, depending on a variety of factors that affect the development of the crystalline structure, such as pressure, intensity of the electric field, controlled melt crystallization, precipitation from different solvents, or seeding crystallization (e.g., surfactants) [74]. The alpha and beta forms are most common in practical situations. Generally, the alpha form is generated in normal melt processing; the beta form develops under mechanical deformation of melt-fabricated specimens. The gamma form arises under special circumstances, and the delta form is obtained by distortion of one of the phases under high electrical fields. The density of PVDF in the alpha crystal form is 1.98 g/cm$^3$; the density of amorphous PVDF is 1.68 g/cm$^3$. Thus the typical density of commercial products in the range from 1.75 to 1.78 g/cm$^3$ reflects a degree of crystallinity around 40% [74].

The structure of PVDF chain, namely alternating $CH_2$ and $CF_2$ groups, has an effect on its properties which combine some of the best performance characteristics of both polyethylene $(-CH_2-CH_2-)_n$ and PTFE $(-CF_2-CF_2-)_n$. Certain commercial grades of PVDF are copolymers of VDF with small amounts (typically less than 6%) of other fluorinated monomers, such as HFP, CTFE, and TFE. These exhibit somewhat different properties than the homopolymer [75].

The unique dielectric properties and polymorphism of PVDF are the source of its high piezoelectric and pyroelectric activity [74]. The relationship between ferroelectric behavior, which includes piezoelectric and pyroelectric phenomena and other electrical properties of the polymorphs of PVDF, is discussed in Ref. [76].

The structure yielding a high dielectric constant and a complex polymorphism also exhibits a high dielectric loss factor. This excludes PVDF from applications as an insulator for conductors of high-frequency currents since the insulation could heat up and possibly even melt. On the other hand, because of that, PVDF can be readily melted by radio frequency or dielectric heating, and this can be utilized for certain fabrication processes or joining [77]. High-energy radiation cross-links PVDF, and the result is the enhancement of mechanical properties. This feature makes it unique among vinylidene polymers, which typically are degraded by high-energy radiation [76]. Otherwise, the mechanical strength of PVDF can be greatly increased by orientation [74].

PVDF exhibits an excellent resistance to inorganic acids, weak bases and halogens, oxidizing agents even at elevated temperatures, and to aliphatic, aromatic, and chlorinated solvents. Strong bases, amines, esters, and ketones cause swelling, softening, and dissolution, depending on conditions [78]. Certain esters and ketones can act as latent solvents for PVDF in dispersions. Such systems solvate the polymer as the temperature is raised during the fusion of the coating, resulting in a cohesive film [79]. PVDF is among the few semicrystalline polymers that exhibit thermodynamic compatibility with other polymers [80], in particular with acrylic and methacrylic resins [81].

#### 2.2.2.7 Structure and Related Properties of THV Fluoroplastics

THV terpolymers have a unique combination of properties that include relatively low processing temperatures, bondability (to itself and other substrates), high flexibility, excellent clarity, low refractive index, and efficient electron-beam cross-linking [82]. It also exhibits properties associated with fluoroplastics, namely very good chemical resistance, weatherability, low friction, and low flammability.

The melting temperatures of the THV commercial products range from 120°C (248°F) to 225°C (437°F) [82]. The lowest melting grade has the lowest chemical resistance and is easily soluble in ketones and ethyl acetate, is the most flexible, and is the easiest to cross-link by electron beam of all grades. On the other hand the highest melting grade has also the highest chemical resistance and resistance to permeation [82].

THV terpolymers can be readily bonded to themselves and many other plastics and elastomers and unlike other fluoroplastics do not require surface treatment, such as chemical etching or corona treatment. However, in some cases, tie layers are required to achieve a good bonding to other materials [83]. They are transparent to a broad band of light (UV to infrared) with an extremely low haze. Its refractive index is very low and depends on the grade [84].

### 2.2.3 Processing of Melt-Processible Fluoroplastics

In general, melt-processible fluoroplastics can be processed like any other thermoplastic polymers by conventional processing methods such as extrusion, injection molding, rotational molding, blow molding, powder and fluidized bed coating, and dipping and coating of different substrates if the given polymer is available in an aqueous dispersion or can be dissolved in a volatile solvent. The need for highly fluorinated thermoplastic polymers that, unlike PTFE, could be fabricated by conventional melt-processing methods led to the development of a group of resins that are copolymers of TFE with other perfluorinated monomers and are referred to as *melt-processible perfluoroplastics*. There are three commercial products, namely, the copolymer of TFE and HFP commonly known as FEP. Copolymerization of TFE with PPVE leads to *PFA resins*, and copolymerization of TFE with perfluoromethylvinyl ether produces *MFA resins*.

#### 2.2.3.1 Processing of Fluorinated Ethylene Propylene Resins

FEP resins are available in low melt viscosity, as extrusion grade, in intermediate viscosity, in high melt viscosity, and as aqueous dispersions [30]. They can be processed by techniques commonly used for thermoplastics, such as extrusion, injection molding, transfer molding, rotational molding, slush molding, and powder and fluidized bed coating [85], and can be expanded into foams [86]. Compression and transfer molding of FEP resins can be done but with some difficulty. Extrusion of FEP is used for primary insulation or cable jackets and for tubing and films. In addition, FEP aqueous dispersions can be used for dipping of fabrics and coating of a variety of substrates.

Processing temperatures used for FEP resins are usually up to 427°C (800.6°F), at which temperatures highly corrosive products are generated. Therefore the parts of the processing equipment that are in contact with the melt must be made of special corrosion-resistant alloys to assure a trouble-free operation.

#### 2.2.3.2 Processing of PFA Resins

PFA can be processed by standard techniques used for thermoplastics, such as extrusion and injection molding and transfer molding at temperatures up to 425°C (797°F). High processing temperatures are required because PFA has a high melt viscosity with activation energy lower than most thermoplastics, 50 kJ/mol [87]. Extrusion and injection molding are done at temperatures typically above 390°C (734°F) and relatively high shear rates. PFA exhibits a sudden transition from the Newtonian behavior to an overflow regime when the critical value of the shear rate is reached. PFA is thermally a very stable polymer, but it still is subject to thermal degradation at processing temperatures, the extent of which depends on temperature, residence time, and the shear rate. Thermal degradation occurs mainly from the end groups; chain scission becomes evident at temperatures above 400°C (752°F) depending on the shear rate. Thermal degradation usually causes discoloration and bubbles [87]. Because at the high processing temperatures large amounts of highly corrosive products are generated, the parts of the equipment have to be made from corrosion-resistant nickel-based alloys to assure a trouble-free operation at temperatures up to 420°C (788°F). PFA can be extruded into films, tubing, rods, and foams [86]. PFA can also be processed as an aqueous dispersion.

#### *2.2.3.3 Processing of MFA Resins*

MFA has the melting temperature 20°C (68°F) lower than PFA, but its processing behavior is similar to that of PFA. As in the case of PFA, standard thermoplastic melt-processing techniques can be used for all grades of MFA. The melt is also corrosive to most metals at the processing temperatures, and processing equipment has to be constructed from corrosion-resistant nickel-based alloys. MFA can also be extruded into films, tubing, rods, and foams and processed as an aqueous dispersion.

#### *2.2.3.4 Processing of ETFE Resins*

ETFE copolymers can be readily fabricated by a variety of melt-processing techniques [88]. They have a wide processing window, in the range 280°C–340°C (536°F–644°F) and can be extruded into films, tubing, and rods or as thin coating on wire and cables.

Welding of ETFE parts can be done easily by spin welding, ultrasonic welding, and conventional butt-welding using flame and ETFE rod. The resins bond readily to untreated metals, but chemical etch corona and flame treatment can be used to increase adhesion further [88].

ETFE resins are very often compounded with varied ingredients (such as fiberglass or bronze powder) or modified during their processing. The most significant modification is cross-linking by peroxides or ionizing radiation. The cross-linking results in improved mechanical properties, higher upper-use temperatures, and a better cut-through resistance without significant sacrifice of electrical properties or chemical resistance [89]. The addition of fillers improves creep resistance, improves friction and wear properties, and increases softening temperature [90]. ETFE can also be processed as an aqueous dispersion; however, at this writing, no ETFE dispersions are commercially available, and reportedly, they were discontinued.

#### 2.2.3.5 Processing of Poly(Chlorotrifluoroethylene) Resins

PCTFE can be processed by most of the techniques used for thermoplastics, mainly extrusion and injection molding. Processing temperatures can be as high as 350°C (662°F) with melt temperatures leaving the nozzle in the range of 280°C–305°C (536°F–581°F). Since relatively high-molecular-weight resins are required for adequate mechanical properties, the melt viscosities are somewhat higher than those usual in the processing of thermoplastics. The reason is a borderline thermal stability of the melt, which does not tolerate sufficiently high processing temperatures [71].

#### 2.2.3.6 Processing of ECTFE Resins

The most common form of ECTFE is hot-cut pellets, which can be used in all melt-processing techniques, such as extrusion, injection molding, blow molding, compression molding, and fiber spinning [72]. ECTFE is corrosive in melt; the surfaces of machinery that come in contact with the polymer must be lined with a highly corrosion-resistant alloy, for example, Hastelloy C-276. Recently developed grades with improved thermal stability and acid scavenging can be processed on conventional equipment [91].

#### 2.2.3.7 Processing of Polyvinylidene Fluoride Resins

PVDF resins for melt processing are supplied as powders or pellets with a rather wide range of melt viscosities. Lower viscosity grades are used for injection molding of complex parts, while the low-viscosity grades have high enough melt strength for the extrusion of profiles, rods, tubing, pipe, film, wire insulation, and monofilament. PVDF extrudes very well, and there is no need to use lubricants or heat stabilizers [78]. The equipment for the melt processing of PVDF is the same as that for PVC or polyolefins, as during normal processing of PVDF no corrosive products are formed. Extrusion temperatures vary between 230°C and 290°C (446°F and 554°F), depending on the equipment and the profile being extruded. Water quenching is used for wire insulation, tubing, and pipe, whereas sheet and cast film from slit dies are cooled on polished steel rolls kept at temperatures between 65°C and 140°C (149°F and 284°F). PVDF films can be monoaxially and biaxially oriented [92].

PVDF resins can be molded by compression, transfer, and injection molding in conventional molding equipment. The mold shrinkage can be as high as 3% due to the semicrystalline nature of PVDF. Molded parts often require annealing at temperatures between 135°C and 150°C (275°F and 302°F) to increase dimensional stability and release internal stresses [93].

Parts from PVDF can be machined, sawed, coined, metallized, and fusion bonded more easily than most other thermoplastics. Fusion bonding usually yields a weld line that is as strong as the part. Adhesive bonding of PVDF parts can be done; epoxy resins produce good bonds [93]. Because of a high dielectric constant and loss factor, PVDF can be readily melted by radio frequency and dielectric heating. This is the basis for some fabrication and joining techniques [77].

PVDF can be coextruded and laminated, but the process has its technical challenges in matching the coefficients of thermal expansion, melt viscosities, and layer adhesion. Special tie layers, often from blends of polymers compatible with PVDF, are used to achieve bonding [94,95].

#### 2.2.3.8 Processing of THV Fluoroplastics

THV Fluoroplastics can be processed by virtually any method used generally for thermoplastics, including extrusion, coextrusion, tandem extrusion, blown film extrusion, blow molding, injection molding, vacuum forming, and as skived film and solvent casting.

Generally, processing temperatures for THV Fluoroplastics are comparable to those used for most thermoplastics. In extrusion, melt temperatures at the die are in the 230°C–250°C (446°F–482°F) range. These relatively low processing temperatures open new options for combinations of different melts (coextrusion, cross-head extrusion, coblow molding) with thermoplastics as well as with various elastomers [96]. Another advantage of the low processing temperatures is that they are generally below the decomposition temperature of the polymer; thus there is no need to protect equipment against corrosion. THV Fluoroplastics were found to be suitable for coextrusion with a variety of materials into multilayer structures [97].

In injection molding, THV Fluoroplastics are processed at lower temperatures than other fluoropolymers, typically at 200°C–300°C (392°F–572°F) with mold temperatures being 60°C–100°C (140°F–212°F). Generally, standard injection molding equipment is used [98]. THV Fluoroplastics can be easily processed by blow molding alone or with polyolefins. The olefin layer provides a structural integrity while THV Fluoroplastics provide chemical resistance and considerably reduced permeation [96].

THV Fluoroplastics can be readily bonded to itself and other plastics and elastomers. It does not require surface treatment, such as chemical etch or corona treatment, to attain good adhesion to other polymers, although in some cases tie layers are necessary. For bonding THV Fluoroplastics to elastomers an adhesion promoter is compounded to the elastomer substrate [83].

### 2.2.4 Applications for Melt-Processible Fluoropolymers

#### *2.2.4.1 Applications for Fluorinated Ethylene Propylene Resins*

The largest proportion of FEP is used in electrical applications, such as hookup wire, interconnecting wire, thermocouple wire, computer wire, and molded parts for electrical and electronic components. Chemical applications include lined tanks, lined pipes and fittings, heat exchangers, overbraided hose, gaskets, component parts of valves, and laboratory ware [99]. Mechanical uses include antistick applications such as conveyor belts and roll covers. FEP film is used in solar-collector windows because of its light weight, excellent weather resistance, high transparency, and easy installation [99]. FEP film is also used for heat sealing of PTFE-coated fabrics, for example, architectural fabric.

#### *2.2.4.2 Applications for Perfluoroalkoxy Resins*

Perfluoroalkoxy resins are fabricated into high-temperature electrical insulation and into components and parts requiring long flex life [100]. Certain grades are used in chemical industry for process equipment, liners, specialty tubing, and molded articles. Other uses are bellows and expansion joints, liners for valves, pipes, pumps, and fittings. Extruded films from PFA and MFA can be used for heat sealing on a variety of PTFE-coated fabrics or can be oriented and used as such for specialized applications [101]. PFA and MFA resins can be processed into injection-molded, blow-molded, and compression-molded components. High-purity grades are used in the semiconductor industry for demanding chemical applications [101]. Coated metal parts can be made by powder coating.

#### *2.2.4.3 Applications for ETFE Resins*

ETFE is used in electrical applications for heat-resistant insulations and jackets of low-voltage power wiring for mass transport systems, for wiring in chemical plants, and for control and instrumentation wiring for utilities [102]. Because ETFE exhibits an excellent cut-through and abrasion resistance, it is used in airframe wire and computer hookup wire. Electrical and electronic components, such as sockets, connectors, and switch components, are made by injection molding [103]. ETFE has excellent mechanical properties; therefore it is used successfully in seal glands, pipe plugs, corrugated tubing, fasteners, and pump vanes [102]. Its radiation resistance is a reason for its use in nuclear industry wiring [90]. The lower density of ETFE provides advantage over perfluoropolymers in aerospace wiring [90]. Because of its excellent chemical resistance, ETFE is used in the chemical industry for valve components, packings, pump impellers, laboratory ware, and battery and instrument components and for oil well, downhole cables [90].

Heat-resistant grades are used for insulation and jackets for heater cables and automotive wiring and for other heavy-wall applications where operating temperatures up to 200°C (392°F) are experienced for short periods of time or where repeated mechanical stress at 150°C (302°F) is encountered [90]. Another use is wiring for high-rise building and skyscraper fire alarm systems. Thin ETFE films are used in greenhouse applications because of their good light transmission, toughness, and resistance to UV radiation [90]. Biaxially oriented films have excellent physical properties and toughness equivalent to polyester films [104].

Injection-molded parts such as electrical connectors and sockets, distillation column plates and packings, valve bodies, pipe and fitting linings are easily made because ETFE exhibits a low shear sensitivity and wide processing window [90].

ETFE can be extruded continuously into tubing, piping, and rod stock. An example of application of extruded tubing is automotive tubing, which takes advantage of its chemical resistance, mechanical strength, and resistance to permeation of hydrocarbons. A high weld factor (more than 90%) is utilized in butt welding of piping and sheet lining of large vessels [90].

ETFE resins in the powder and bead form are rotationally molded into varied structures, such as pump bodies, tanks, and fittings and linings, mostly for the chemical-process industries. Inserts can be incorporated to provide attachment points or reinforcement [90]. Adhesion to steel, copper, and aluminum can be up to 3 kN/m (5.7 pli) peel force [103].

Carbon-filled ETFE resins (about 20% carbon) exhibit antistatic dissipation and are used in self-limiting heater cables and other antistatic or semiconductive applications [90].

Certain grades of ETFE are used for extruded foams with void contents from 20% to 50%. The closed foam cells are 0.001–0.003 in. (0.02–0.08 mm) in diameter. Special grades of ETFE processed in gas-injection foaming process may have void contents up to 70%. Foamed ETFE is used in electrical applications, mainly in cables, because it exhibits lower apparent dielectric constant and dissipation factor and reduces cable weight [52].

#### 2.2.4.4 Applications for Poly(Chlorotrifluoroethylene) Resins

Although PCTFE can be processed by the most common melt-processing methods, the largest volume is used for the extrusion of specialty films for packaging in applications where there are high moisture barrier demands, such as pharmaceutical blister packaging and health-care markets. In electroluminescent (EL) lamps PCTFE film is used to encapsulate phosphor coatings, which provide an area light when electrically excited [105]. The film acts as a water vapor barrier protecting the moisture-sensitive phosphor chemicals. EL lamps are used in aircraft, military, aerospace, automotive, business equipment applications, and in buildings. Another use for PCTFE films is for packaging of corrosion-sensitive military and electronic components. Because of excellent electrical insulation properties, these films can be used to protect sensitive electronic components, which may be exposed to humid or harsh environment. They can be thermoformed to conform to any shape and detail. PCTFE films are also used to protect the moisture-sensitive liquid crystal display panels of portable computers [106].

PCTFE films can be laminated to a variety of substrates, such as PVC, PETG, amorphous polyethylene terephthalate, or polypropylene. Metallized films are used for electronic dissipative and moisture barrier bags for sensitive electronic components, for packaging of drugs, and for medical devices. Other applications for PCTFE are in pump parts, transparent sight glasses, flowmeters, tubes, and linings in the chemical industry and for laboratory ware [107].

#### 2.2.4.5 Applications for ECTFE Resins

The single largest application for ECTFE has been as primary insulation and jacketing [108] for voice and copper cables used in building plenums [72]. In automotive applications, ECTFE is used for jackets of cables inside fuel tanks for level sensors, for hookup wires, and in heating cables for car seats. Chemically foamed ECTFE is used in some cable constructions [109].

In the chemical-process industry, it is often used in chlorine/caustic environment in cell covers, outlet boxes, lined pipes, and tanks.

In the pulp and paper industries, pipes and scrubbers for bleaching agents are lined with ECTFE. Powder-coated tanks, ducts, and other components find use in semiconductor and chemical-process industries. Monofilament made from ECTFE is used for chemical-resistant filters and screens [107].

Other applications include rotomolded tanks and containers for the storage of corrosive chemicals, such as nitric or hydrochloric acid. Extruded sheets can be thermoformed into various parts, such as battery cases for heart pacemakers [107]. ECTFE film is used as release sheet in the fabrication of high-temperature composites for aerospace applications. Braided cable jackets made from monofilament strands are used in military and commercial aircraft as a protective sleeve for cables [110].

#### 2.2.4.6 Applications for Polyvinylidene Fluoride Resins

PVDF is widely used in the chemical industry in fluid-handling systems for solid and lined pipes, fittings, valves, pumps, tower packing, tank liners, and woven filter cloth. Because it is approved by the Federal Drug Administration for food contact, it can be used for fluid-handling equipment and filters in the food, pharmaceutical, and biochemical industries. It also meets high standards for purity, required in the manufacture of semiconductors and, therefore, is used for fluid-handling systems in the semiconductor industry [111]. PVDF is also used for the manufacture of microporous and ultrafiltration membranes [112,113].

In electrical and electronic industries, PVDF is used as a primary insulator on computer hookup wire. Irradiated (cross-linked) PVDF jackets are used for industrial control wiring [114] and self-limiting heat-tracing tapes used for controlling the temperature of process equipment as well as ordnance [115,116]. Extruded and irradiated heat-shrinkable tubing is used to produce termination devices for aircraft and electronic equipment [117]. Because of its very high dielectric constant and dielectric loss factor, the use of PVDF insulation is limited to only low-frequency conductors. Under certain conditions, PVDF films become piezoelectric and pyroelectric. The piezoelectric properties are utilized in soundproof telephone headset, infrared sensing, a respiration monitor, a high-fidelity electric violin, hydrophones, keyboards, and printers [118].

#### 2.2.4.7 Applications for THV Fluoroplastics

Because THV Fluoroplastics are highly flexible, resistant to chemicals, and automotive fuels and have good barrier properties, they are used as a permeation barrier in various types of flexible hose in automotive applications and in the chemical-process industry. Liners and tubing can be made from an electrostatic dissipative grade of THV, which is sometimes required for certain automotive applications [119].

THV is used for wire and cable jacketing, which is often cross-linked by electron beam to improve its strength and increase its softening temperature. It is also used as primary insulation in less demanding applications, where high flexibility is required [120].

The low refractive index of THV (typically 1.355) is utilized in light tubes and communication optical fiber applications where high flexibility is required. Its optical clarity and impact resistance make it suitable as film for laminated safety glass for vehicles and for windows and doors in psychiatric and correctional institutions. An additional advantage is that the film does not burn or support combustion, which may be a major concern in some applications [121].

Other applications for THV are flexible liners (drop-in liners or bag liners), used in chemical-process industries and other industries, and blow-molded containers, where it enhances the resistance to permeation when combined with a less expensive plastic (e.g., high-density polyethylene), which provides the structural integrity [120]. Optical clarity, excellent weatherability, and flexibility make THV suitable as a protection of solar cell surface in solar modules [122].

## 2.3 Other Thermoplastic Fluoropolymers

### 2.3.1 Industrial Processes for the Production of Other Thermoplastic Fluoropolymers

This section covers a group of thermoplastic fluoropolymers, which are different in properties, processing, and applications, although all are used for the manufacture either of films or coatings, namely PVF, fluorinated thermoplastic elastomers, and amorphous fluoroplastics.

#### 2.3.1.1 Industrial Process for the Production of Polyvinyl Fluoride

Vinyl fluoride is polymerized by free radical processes as most common commercial fluoropolymers, but it is more difficult to polymerize than TFE or VDF and requires higher pressures [123]. The temperature range for the polymerization in aqueous media is reported as being from 50°C to 150°C (122°F to 302°F) and pressures range from 3.4 to 34.4 MPa (500 to 5000 psi). Catalysts for the polymerization are peroxides and azo-compounds [124]. Continuous process, also in aqueous media, is carried out at a temperature of 100°C (212°F) and pressure of 27.5 MPa (4000 psi) [125]. The use of perfluoroalkylpropyl amine salts as emulsifiers in the aqueous polymerization enhances the polymerization rate and yield and produces a polymer with an excellent color [126]. The polymerization temperature has influence on the crystallinity and the melting point of the resulting polymer. Higher temperatures increase branching [124]. PVF characterization as a resin has not been published. The DuPont Company produces and markets only PVF films (Tedlar), which are available as clear, colored, filled, and oriented products. More on the subject in Chapter 5, Films From Polyvinyl Fluoride.

#### 2.3.1.2 Industrial Processes for Production of Fluorinated Thermoplastic Elastomers

Thermoplastic fluoroelastomers (TPEs) are unique polymeric materials exhibiting elastic behavior similar to cross-linked rubber but can be processed by conventional thermoplastics methods without curing (cross-linking). This allows flash from molding and other scrap as well as postconsumer waste to be recovered and reused. They are essentially phase-separated systems [127]. Usually one phase is hard and solid at the ambient temperature and the other one is soft and elastic. The hard phase forms the *physical cross-links*, which are thermoreversible. Often the phases are bonded chemically by block or graft polymerization. Such materials are most frequently referred to as A−B−A block copolymers made often by *living radical copolymerization*. Another major group of thermoplastic elastomers are thermoplastic vulcanizates (TPVs) prepared by *dynamic vulcanization*. The products display a disperse morphology, where particles of cross-linked elastomer are dispersed in thermoplastic matrix [128]. More specifics on production of fluorinated thermoplastic elastomers (FTPEs) are in Section 2.3.2.2.

#### 2.3.1.3 Industrial Processes for the Production of Amorphous Perfluoropolymers

Perfluoropolymers based on copolymers of 2,2-bistrifluoromethyl-4,5-difluoro-1,3-dioxole (PDD) was developed by DuPont under the trade name Teflon AF [129].

PDD readily copolymerizes with TFE and other monomers containing fluorine, such as VDF, CTFE, vinyl fluoride, and PVE via free radical copolymerization, which can be carried out in either aqueous or nonaqueous media. It also forms an amorphous homopolymer with a $T_g$ of 335°C (635°F) [130]. Specifically the commercial product Teflon AF is produced by aqueous copolymerization of PDD and TFE using fluorosurfactants and ammonium persulfate or other metal persulfate initiators [130].

Another pefluoropolymer of this type has been developed by Asahi Glass and is available on the market under the trade name CYTOP. This polymer is prepared by cyclopolymerization of perfluorodiene [131]. More details on this product are in Section 2.3.2.3.

### 2.3.2 Structure and Related Properties of Other Thermoplastic Fluoropolymers

This section covers structures and properties of a group of other thermoplastic fluoropolymers, each of them rather different that the others, namely PVF, FTPEs, and of amorphous perfluoropolymers.

#### 2.3.2.1 Structure and Related Properties of Polyvinyl Fluoride

Although PVF resembles PVC in its low water absorption, resistance to hydrolysis, insolubility in common solvents at room temperature, and a tendency to split off hydrogen halides at elevated temperatures, it has a much greater tendency to crystallize.

PVF exhibits excellent resistance to weathering, outstanding mechanical properties, and inertness toward a wide variety of chemicals, solvents, and staining agents, excellent hydrolytic stability, and high dielectric strength and dielectric constant [132–134]. Films of PVF retain their form and strength even when boiled in strong acids and bases. At ordinary temperatures the film is not affected by many classes of common solvents, including hydrocarbon and chlorinated solvents. It is partially soluble in a few highly polar solvents above 149°C (300.2°F). It is impermeable to greases and oils. Clear PVF films are essentially transparent to solar radiation in the near UV, visible, and near-infrared regions of the spectrum.

PVF films exhibit an outstanding resistance to solar degradation. Unsupported transparent PVF films have retained at least 50% of their tensile strength after 10 years in Florida facing south at 45 degrees. Pigmented films properly laminated to a variety of substrates impart a long service life. Most colors exhibit no more than five NBS-unit (Modified Adams Color Coordinates) color change after 20 years vertical outdoor exposure. Additional protection of various substrates against UV attack can be achieved with UV-absorbing PVF films. PVF films exhibit high dielectric constant and a high dielectric strength [133]. More on the subject in Chapter 5, Films From Polyvinyl Fluoride.

#### 2.3.2.2 Structure and Related Properties of Fluorinated Thermoplastic Elastomers

FTPEs are represented by several types of materials. Daikin developed, manufactures, and is marketing **A**–**B**–**A** block copolymers made by the semibatch emulsion process using fluorocarbon diiodide transfer [135]. The center *elastomeric* **B** block soft segments are made in a first polymerization step. After removal of monomers and recharging a different monomer composition, the plastic **A** block *hard* segments are polymerized on the ends of the **B** blocks [136]. The main commercial product is DAI-EL T-530. The basic patent requires that hard segments have molecular weight of at least 10,000, corresponding to a degree of polymerization (DP) of at least 140 U, sufficient for crystallization with melting point about 220°C (428°F). Central soft blocks would then have molecular weight at least 110,000, with DP is 110 U or more. The high fluorine content of the soft blocks gives the product excellent fluid resistance and a glass transition temperature of about −8°C (17.6°F). The thermoplastic can be extruded and formed at temperatures above the melting range; after cooling, crystallization of the hard segments gives parts with good dimensional stability at temperatures up to about 120°C (248°F). Typical applications include tubing, sheet, O-rings, and molded parts. Daikin has also developed and is marketing a fluorinated TPVs (FTPVs), with the trade name DAI-EL FluoroTPV [137]. The resistance to automotive automatic transmission oil in comparison to copolymer FKM and FEPM (cured fluorocarbon elastomers). It can be readily processed by extrusion with processing temperatures in the range from 220°C to 260°C. FTPEs became attractive to automotive seal suppliers, since scrap loss can be reduced greatly, compared to thermoset elastomers. Some improvement in high-temperature performance may be necessary, however.

Another product is a graft copolymer type comprising main-chain fluoroelastomers and side-chain fluoroplastics. This type is offered by Central Glass Co. under the trade name Cefral Soft [138].

FTPVs have been also developed by Freudenberg NOK-GP and offered as FluoroXprene [139]. They are essentially dynamically vulcanized blends of fluorocarbon elastomers and fluoroplastics, such as PVDF, ETFE, ECTFE, THV, FEP, and MFA prepared in either batch or continuous process. The continuous process using a twin-screw extruder is preferred. The morphology is that typical for a TPV, that is, dispersed cross-linked FKM in the fluoroplastic matrix [140]. The fluid resistance is considerably better than that of FKM, mainly because the semicrystalline fluoroplastic matrix protects the elastomeric particles. The FTPV exhibits fuel permeation resistance superior to that of FKM materials. The FTPVs containing VDF and ethylene monomeric units can be cross-linked by ionizing radiation (electron beam, gamma rays, X-rays) if desired. Among other things, it is compatible with several elastomers and is very flexible. Because of its unique properties, it can be used in combination with NBR in automotive industry for multilayer fuel hose combined with NBR and for filler neck hoses as well as O-rings of different sizes. At this writing, there were no reports of the use of these FTPEs for films, although because of their properties, they could be.

#### 2.3.2.3 Structure and Related Properties of Amorphous Perfluoropolymers

Perfluoropolymers based on copolymers of 2,2-bistrifluoromethyl-4,5-difluoro-1,3-dioxole (PDD) was developed by DuPont under the trade name Teflon AF (see Section 2.3.1.3). These products retain the outstanding chemical, thermal, and surface properties associated with perfluorinated polymers and exhibit unique electrical, optical, and solubility characteristics at the same time [129,130].

Teflon AF copolymers have a perfluorinated structure as do PTFE, PFA, and FEP, and therefore they exhibit similar high-temperature stability, chemical resistance, low surface energy, and low water absorption. Unlike PTFE, PFA, and FEP, which are semicrystalline, the completely amorphous Teflon AF copolymers differ considerably in that they are soluble in several perfluorinated solvents at room temperature and have high optical transmission across a broad wavelength region from UV to near infrared [141]. Other differences are lower refractive indexes and dielectric constants and high gas permeability. The refractive index of Teflon copolymers is the lowest known for any solid organic polymer (the respective values for Teflon AF 1600 and AF 2400 are 1.31 and 1.29 at 20°C at the sodium line) and depends on the glass transition temperature of the polymer. The presence of the dioxole structure in the chain imparts a higher stiffness and high tensile modulus [141].

The other pefluoropolymer of this type, CYTOP, offered by Asahi Glass is prepared by cyclopolymerization of perfluorodiene [142] (see Section 2.3.1.3). CYTOP has physical and chemical properties similar to PTFE and PFA. Its tensile strength and yield strength are higher than those of PTFE and PFA. It also has unique optical properties: its films are transparent in the range from 200 to 700 nm, and its clarity is very high even in the UV region. CYTOP is soluble in selected fluorinated solvents. Such solutions have a very low surface tension, which allows them to be spread onto porous materials and to cover the entire surface. Films without pinholes and of uniform thickness can be prepared from solutions [142].

## 2.3.3 Processing of Other Thermoplastic Fluoropolymers

### *2.3.3.1 Processing of Polyvinyl Fluoride*

PVF is considered a thermoplastic, but it cannot be processed by conventional thermoplastic techniques, because it is unstable above its melting point. However, it can be fabricated into self-supporting films and coatings by using latent solvents (see below) [143]. It can be compression molded, but this method is not commonly used [144]. Because of a large number of hydrogen bonds and a high degree of crystallinity, PVF is insoluble at room temperature. However, some highly polar latent solvents, such as propylene carbonate, dimethylformamide, dimethyl acetamide, butyrolactone, and dimethyl sulfoxide, dissolve it above 100°C (212°F) [143]. The use of latent solvents is the basis of processes to manufacture films and coatings. A latent solvent suitable for that has to have the appropriate volatility to allow the polymer particles to coalesce before complete evaporation.

Structurally modified PVF has been extruded [145]. Thin films are manufactured by extrusion of a dispersion of PVF in a latent solvent [146]. Such dispersion contains usually pigments, stabilizers, plasticizers, and flame retardants as well as deglossing agents if needed. The solvent is removed by evaporation. The extruded film can be biaxially oriented, and the solvent is removed by evaporation only after the orientation is completed.

Homopolymers and copolymers of vinyl fluoride can be applied to substrates as dispersion in a latent solvent or water or by powder coating. Usually, the substrate does not need to be primed [147]. The dispersions may be applied by spraying, reverse roll coating, dip coating, or centrifugal casting. Another method is dipping a hot article into the dispersion below 100°C (212°F). PVF films are most frequently produced by casting on a continuous moving belt. More details on the process are provided in Chapter 5, Films From Polyvinyl Fluoride.

PVF films often require a surface treatment to improve bonding to other materials. Among these are flame treatment [148], electric discharge [149], chemical etching, and plasma treatment [150].

#### 2.3.3.2 Processing of Fluorinated Thermoplastic Elastomers

Currently, there are several types of FTPEs (see Section 2.3.2.2), each exhibiting different structure and processing properties. In general, they can be processed by conventional thermoplastic melt-processing techniques, including extrusion, injection molding, transfer molding, blow molding, and compression molding. The processing temperatures depend on the melting temperature of the hard segment of the molecule in the case of multiblock types of copolymers or on the melting temperature of the fluoroplastic matrix in the dynamically vulcanized systems, known as *TPVs*. Another important factor is the overall rheological properties of the system.

#### 2.3.3.3 Processing of Amorphous Perfluoropolymers

Amorphous perfluoropolymers are processed by conventional melt-processing techniques, such as extrusion, injection molding, compression molding [151], and solution processing, including spin coating, dip coating, spraying, and potting [152]. Special techniques for producing very thin films are laser ablation and vacuum pyrolysis [153].

## 2.3.4 Applications for Other Thermoplastic Fluoropolymers

### 2.3.4.1 Applications for Polyvinyl Fluoride

PVF is almost exclusively used as film for lamination with a large variety of substrates. Its main function is as a protective and decorative coating. PVF films can be made transparent or pigmented to a variety of colors and can be laminated to hardboard, paper, flexible PVC, polystyrene, rubber, polyurethane, and other substrates [107]. These laminates are used for wall coverings, aircraft cabin interiors, pipe covering, duct liners, etc. For covering metal and rigid PVC the film is first laminated to flat, continuous metal or vinyl sheets using special adhesives, and then the laminate is formed into the desired shapes. The laminates are used for exterior sidings of industrial and residential buildings. Other applications are highway sound barriers [154] automobile truck and trailer siding [155], vinyl awnings, and backlit signs [156]. On metal or plastic, PVF surfaces serve as a primer coat for painting or adhesive joints [157]. PVF films are used as a release sheet for bag molding of composites from epoxide, polyester, and phenolic resins and in the manufacture of circuit boards [158]. Other uses of PVF films are in greenhouses, flat-plate solar collectors, and in photovoltaic cells. Dispersions of PVF are used for coating the exterior of steel hydraulic brake tubing for corrosion protection [157]. More on the subject in Chapter 5, Films From Polyvinyl Fluoride.

### 2.3.4.2 Applications for Fluorinated Thermoplastic Elastomers

The most common applications for FTPEs are in chemical and semiconductor industries (O-rings, V-rings, gaskets, and diaphragms) because of their excellent chemical resistance and purity [159]. These parts are often cross-linked by ionizing radiation [160]. Other parts for these industries are tubing, liners of multilayer hoses for corrosive gases or ultrapure water, and liners for vessels for inorganic acids (e.g., HF) [161]. Other uses include as wire coatings and sheeting and coating of wires [162,163], tents and greenhouses, tubing, bottles and packaging in food processing, and in sanitary goods [164].

#### 2.3.4.3 Applications for Amorphous Perfluoropolymers

Teflon AF type amorphous perfluoropolymers have many uses, including antireflecting coatings [165], low dielectric coatings, deep UV pellicles used in electronic chip manufacturing processes [166], cladding of plastic optical fibers [167], and as a low dielectric insulator for high-performance interconnects [168].

The CYTOP type amorphous perfluoropolymers are used in pellicles, which are photomask covers for dust protection sand to prevent contamination during the microlithography process in semiconductor production [169]. They are also used in antireflective coatings because of their low refractive index as well as in cladding of PMMA optical fibers [170].

## References

[1] S. Ebnesajjad, Fluoroplastics, Vol. 1, Non-Melt Processable Fluoroplastics, second ed., William Andrew Publishing, Norwich, NY, 2000, p. 96.
[2] S. Ebnesajjad, Fluoroplastics, Vol. 1, Non-Melt Processable Fluoroplastics, second ed., William Andrew Publishing, Norwich, NY, 2000, p. 98.
[3] S. Sheratt, Encyclopedia of Chemical Technology, vol. 9, John Wiley & Sons, New York, 1966, p. 813.
[4] S.G. Bankoff, U.S. Patent 2,612,484, E.I. du Pont de Nemours and Co., 1952.
[5] S. Sheratt, Encyclopedia of Chemical Technology, vol. 9, John Wiley & Sons, New York, 1966, p. 814.
[6] J.G. Drobny, Technology of Fluoropolymers, second ed., CRC Press, Boca Raton, FL, 2009, p. 11.
[7] S.V. Gangal, U.S. Patent 4,342,675, E.I. du Pont de Nemours and Co., 1982.
[8] K.L. Berry, US Patent 2,478,229, E.I. du Pont de Nemours and Co., 1949.
[9] G.P. Koo, in: L.A. Wall (Ed.), Fluoropolymers, John Wiley & Sons, New York, 1972, p. 5010.
[10] J.G. Drobny, Technology of Fluoropolymers, second ed., CRC Press, Boca Raton, FL, 2009, p. 29.
[11] S. Sheratt, Encyclopedia of Chemical Technology, vol. 9, John Wiley & Sons, New York, 1966, p. 817.
[12] J.G. Drobny, Technology of Fluoropolymers, second ed., CRC Press, Boca Raton, FL, 2009, p. 174.

[13] J.G. Drobny, Technology of Fluoropolymers, second ed., CRC Press, Boca Raton, FL, 2009, p. 38.
[14] K. Hintzer, G. Löhr, in: J. Scheirs (Ed.), Modern Fluoropolymers, John Wiley & Sons, New York, 1997, p. 239.
[15] K. Hintzer, G. Löhr, in: J. Scheirs (Ed.), Modern Fluoropolymers, John Wiley & Sons, New York, 1997, p. 251.
[16] J.G. Drobny, Technology of Fluoropolymers, second ed., CRC Press, Boca Raton, FL, 2009, p. 133.
[17] D.I. Mc Cane, in: N.M. Bikales (Ed.), Encyclopedia of Polymer Science and Technology, John Wiley & Sons, New York, 1970, p. 623.
[18] C.A. Sperati, in: I.I. Rubin (Ed.), Handbook of Plastic Materials and Technology, John Wiley& Sons, New York, 1990, p. 121.
[19] R.W. Gore, US Patent 3,962,153, W.L. Gore & Associates, 1976.
[20] Y. Hishinuma, T. Chikahisa, H. Yoshikawa 1998 Fuel Cell Seminar, Nov. 16–19, 1998, Palm Springs Convention Center, Palm Springs, CA, 1998, p. 655.
[21] A. El-Sayed, et al., J. Appl. Polym. Sci. 38 (1989) 1229.
[22] J.R. Hallahan, et al., J. Appl. Polym. Sci. 13 (1969) 807.
[23] E.H. Cirlin, D.H. Kaelble, J. Polym. Sci., Polym. Phys. Ed. 11 (1973) 785.
[24] R. Baumhard-Neto, et al., J. Polym. Sci., Polym. Chem. Ed. 19 (1981) 819.
[25] C.A. Rappaport, US Patent 2,809,130, General Motor Corporation, 1957.
[26] R.C. Doban, US Patent 2,871,144, E.I. du Pont de Nemours & Co., 1959.
[27] W.T. Miller, US Patent 2,598,283, U.S. Atomic Energy Commission, 1952.
[28] R.A. Naberezhnykh, et al., Dokl. Akad. Nauk SSSR 214 (1974) 149.
[29] A.S. Kabankin, et al., Vysokomol. Soed., Ser. A 12 (1970) 267.
[30] S.V. Gangal, in: H.F. Mark, J.I. Kroschwitz (Eds.), Encyclopedia of Polymer Science and Technology, vol. 16, John Wiley & Sons, New York, 1989, p. 603.
[31] S.V. Gangal, in: third ed., M. Grayson (Ed.), Encyclopedia of Chemical Technology, vol. 11, John Wiley & Sons, New York, 1978, p. 24.
[32] J.F. Harris Jr., D.I. McCrane, US Patent 3,132,123, DuPont Co., 1964.
[33] D.P. Carlson, US Patent 3,536,733, DuPont Co., 1997.
[34] W.F. Gresham, A.F. Vogelpohl, US Patent 3,635,926, E.I. du Pont de Nemours & Co., 1976.
[35] C.A. Sperati, in: I.I. Rubin (Ed.), Handbook of Plastic Materials and Technology, John Wiley& Sons, New York, 1990, p. 87.
[36] D.L. Kerbow, in: J. Scheirs (Ed.), Modern Fluoropolymers, John Wiley & Sons, Ltd, Chichester, 1997, p. 302.
[37] S. Ebnesajjad, Fluoroplastics, Volume 2: Melt Processable Fluoroplastics, second ed., Elsevier, Oxford, 2015, p. 178.
[38] H.S. Booth, P.E. Burchfield, J. Am. Chem. Soc. 55 (1933) 2231.

[39] D.P. Carlson, W. Schmiegel, in: W. Gerhartz (Ed.), Ullmann's Encyclopedia of Industrial Chemistry, vol. A 11, VCH Publishers, Weinheim, 1988, p. 411.
[40] H.K. Reimschuessel, J. Marti, N.S. Murthy, J. Polym. Sci., A: Polym. Chem. 26 (1988) 43.
[41] S. Schulze, US Patent 4,053,445, 1977.
[42] G. Stanitis, in: J. Scheirs (Ed.), Modern Fluoropolymers, John Wiley & Sons, Chichester, 1997, p. 528.
[43] S. Ebnesajjad, Fluoroplastics, Volume 2: Melt Processable Fluoroplastics, second ed., Elsevier, Oxford, 2015, p. 191.
[44] D.A. Seiler, in: J. Scheirs (Ed.), Modern Fluoropolymers, John Wiley & Sons, Ltd, Chichester, 1997, p. 489.
[45] J.E. Dohany, J.S. Humphrey, in: H.F. Mark, J.I. Kroschwitz (Eds.), Encyclopedia of Polymer Science and Technology, vol. 17, John Wiley & Sons, New York, 1989, p. 534.
[46] T. Locheed, Chem. Process. (1993) 37–38.
[47] D.A. Seiler, in: J. Scheirs (Ed.), Modern Fluoropolymers, John Wiley & Sons, Chichester, 1997, p. 490.
[48] D.P. Carlson, W. Schmiegel, in: W. Gerhartz (Ed.), Ullmann's Encyclopedia of Industrial Chemistry, vol. A11, VCH Publishers, Weinheim, 1988, p. 414.
[49] D.E. Hull, B.V. Johnson, I.P. Rodricks, J.B. Staley, in: J. Scheirs (Ed.), Modern Fluoropolymers, John Wiley & Sons, Chichester, 1997, p. 257.
[50] Teflon® FEP Fluorocarbon Film, Bulletin 5A, Optical, E.I. du Pont de Nemours & Company.
[51] J.G. Drobny, in: M. Gilbert (Ed.), Brydson's Plastics Materials, eighth ed., Elsevier, Oxford, 2017, p. 398.
[52] J.G. Drobny, Technology of Fluoropolymers, second ed., CRC Press, Boca Raton, FL, 2009, p. 81.
[53] S.V. Gangal, in: H.F. Mark, J.I. Kroschwitz (Eds.), Encyclopedia of Polymer Science and Technology, vol. 16, John Wiley & Sons, New York, 1989, p. 604.
[54] S.V. Gangal, in: third ed., M. Grayson (Ed.), Encyclopedia of Chemical Technology, vol. 11, John Wiley & Sons, New York, 1978, p. 29.
[55] D.L. Ryan, British Patent 897,466, 1962.
[56] M. Pozzoli, G. Vitta, V. Arcella, in: J. Scheirs (Ed.), Modern Fluoropolymers, John Wiley & Sons, Chichester, 1997, p. 380.
[57] K. Hintzer, G. Löhr, in: J. Scheirs (Ed.), Modern Fluoropolymers, John Wiley & Sons, New York, 1997, p. 230.
[58] K. Hintzer, G. Löhr, in: J. Scheirs (Ed.), Modern Fluoropolymers, John Wiley & Sons, New York, 1997, p. 233.
[59] M. Pozzoli, G. Vitta, V. Arcella, in: J. Scheirs (Ed.), Modern Fluoropolymers, John Wiley & Sons, Chichester, 1997, p. 384.

[60] M. Pozzoli, G. Vitta, V. Arcella, in: J. Scheirs (Ed.), Modern Fluoropolymers, John Wiley & Sons, Chichester, 1997, p. 385.
[61] M. Pozzoli, G. Vitta, V. Arcella, in: J. Scheirs (Ed.), Modern Fluoropolymers, John Wiley & Sons, Chichester, 1997, p. 386.
[62] M. Pozzoli, G. Vitta, V. Arcella, in: J. Scheirs (Ed.), Modern Fluoropolymers, John Wiley & Sons, Chichester, 1997, p. 387.
[63] J.G. Drobny, Technology of Fluoropolymers, second ed., CRC Press, Boca Raton, FL, 2009, p. 43.
[64] D.L. Kerbow, in: J. Scheirs (Ed.), Modern Fluoropolymers, John Wiley & Sons, Ltd, Chichester, 1997, p. 305.
[65] A.J. Gotcher, P. Gameraad, US Patent 4,155,823, 1979.
[66] TEFZEL® Chemical Use Temperature Guide, Bulletin E-18663, E.I. du Pont de Nemours & Company, 1990.
[67] D.L. Kerbow, in: J. Scheirs (Ed.), Modern Fluoropolymers, John Wiley & Sons, Ltd, Chichester, 1997, p. 304.
[68] C.A. Sperati, in: I.I. Rubin (Ed.), Handbook of Plastic Materials and Technology, John Wiley& Sons, New York, 1990, p. 102.
[69] J.G. Drobny, in: M. Gilbert (Ed.), Brydson's Plastics Materials, eighth ed., Elsevier, Oxford, 2017, p. 400.
[70] C.A. Sperati, in: I.I. Rubin (Ed.), Handbook of Plastic Materials and Technology, John Wiley& Sons, New York, 1990, p. 106.
[71] C.A. Sperati, in: I.I. Rubin (Ed.), Handbook of Plastic Materials and Technology, John Wiley& Sons, New York, 1990, p. 103.
[72] G. Stanitis, in: J. Scheirs (Ed.), Modern Fluoropolymers, John Wiley & Sons, Chichester, 1997, p. 533.
[73] J.G. Drobny, Technology of Fluoropolymers, second ed., CRC Press, Boca Raton, FL, 2009, p. 48.
[74] J.E. Dohany, J.S. Humphrey, in: H.F. Mark, J.I. Kroschwitz (Eds.), Encyclopedia of Polymer Science and Technology, vol. 17, John Wiley & Sons, New York, 1989, p. 536.
[75] J.G. Drobny, Technology of Fluoropolymers, second ed., CRC Press, Boca Raton, FL, 2009, p. 45.
[76] J.E. Dohany, J.S. Humphrey, in: H.F. Mark, J.I. Kroschwitz (Eds.), Encyclopedia of Polymer Science and Technology, vol. 17, John Wiley & Sons, New York, 1989, p. 539.
[77] K. Nakagawa, M. Amano, Polym. Comm. 27 (1986) 310.
[78] J.E. Dohany, J.S. Humphrey, in: H.F. Mark, J.I. Kroschwitz (Eds.), Encyclopedia of Polymer Science and Engineering, vol. 17, John Wiley & Sons, New York, 1989, p. 540.
[79] J.E. Dohany, H. Stefanou, Am. Chem. Soc. Div. Org. Coat. Plast. Chem. Pap. 35 (2) (1975) 83.
[80] D.R. Paul, J.W. Barlow, J. Macromol. Sci. Rev. Macromol. C18 (1980) 109.
[81] J. Mijovic, H.L. Luo, C.D. Han, Polym. Eng. Sci. 22 (4) (1982) 234.

[82] D.E. Hull, B.V. Johnson, I.P. Rodricks, J.B. Staley, in: J. Scheirs (Ed.), Modern Fluoropolymers, John Wiley & Sons, Chichester, 1997, p. 258.
[83] D.E. Hull, B.V. Johnson, I.P. Rodricks, J.B. Staley, in: J. Scheirs (Ed.), Modern Fluoropolymers, John Wiley & Sons, Chichester, 1997, p. 260.
[84] D.E. Hull, B.V. Johnson, I.P. Rodricks, J.B. Staley, in: J. Scheirs (Ed.), Modern Fluoropolymers, John Wiley & Sons, Chichester, 1997, p. 261.
[85] C.A. Sperati, in: I.I. Rubin (Ed.), Handbook of Plastics Materials and Technology, John Wiley & Sons, New York, 1990, p. 96.
[86] C.V. Gupta, (Chapter 15) in: K.C. Frisch, D. Klempner (Eds.), Polymeric Foams, Hanser Publishers, Munich, 1991, p. 349.
[87] K. Hintzer, G. Löhr, in: J. Scheirs (Ed.), Modern Fluoropolymers, John Wiley & Sons, New York; Chichester, p. 234.
[88] E.I. du Pont de Nemours & Co., Extrusion Guide for Melt Processible Fluoropolymers, Bulletin E-85783, E.I. du Pont de Nemours & Co., Wilmington, DE.
[89] D.L. Kerbow, in: J. Scheirs (Ed.), Modern Fluoropolymers, John Wiley & Sons, Chichester, 1997, p. 307.
[90] D.L. Kerbow, in: J. Scheirs (Ed.), Modern Fluoropolymers, John Wiley & Sons, Chichester, 1997, p. 308.
[91] G. Stanitis, in: J. Scheirs (Ed.), Modern Fluoropolymers, John Wiley & Sons, Chichester, 1997, p. 529.
[92] J.G. Drobny, (Chapter 4) Technology of Fluoropolymers, second ed., CRC Press, Boca Raton, FL, 2009.
[93] J.E. Dohany, J.S. Humphrey, in: H.F. Mark, J.I. Kroschwitz (Eds.), Encyclopedia of Polymer Science and Engineering, vol. 17, John Wiley & Sons, New York, 1989, p. 541.
[94] A. Strassel, US Patent 4,317,860, 1982.
[95] N.A. Kitigawa, et al., US. Patent 4,563,393, 1986.
[96] D.E. Hull, B.V. Johnson, I.P. Rodricks, J.B. Staley, in: J. Scheirs (Ed.), Modern Fluoropolymers, John Wiley & Sons, Chichester, 1997, p. 262.
[97] C. Lavallée, The 2nd International Fluoropolymers Symposium, SPI, Expanding Fluoropolymer Processing Options, 1995.
[98] D.E. Hull, B.V. Johnson, I.P. Rodricks, J.B. Staley, in: J. Scheirs (Ed.), Modern Fluoropolymers, John Wiley & Sons, Chichester, 1997, p. 263.
[99] S.V. Gangal, in: H.F. Mark, J.I. Kroschwitz (Eds.), Encyclopedia of Polymer Science and Technology, vol. 16, John Wiley & Sons, New York, 1989, p. 611.
[100] S.V. Gangal, in: H.F. Mark, J.I. Kroschwitz (Eds.), Encyclopedia of Polymer Science and Technology, vol. 16, John Wiley & Sons, New York, 1989, p. 625.
[101] D.P. Carlson, W. Schmiegel, in: W. Gerhartz (Ed.), Ullmann's Encyclopedia of Industrial Chemistry, vol. A11, VCH Publishers, Weinheim, 1988, p. 408.

[102] S.V. Gangal, in: H.F. Mark, J.I. Kroschwitz (Eds.), Encyclopedia of Polymer Science and Technology, vol. 16, John Wiley & Sons, New York, 1989, p. 641.
[103] Tefzel® Fluoropolymers, Product Information, Du Pont Materials for Wire and Cable, Bulletin E-81467, E.I. du Pont de Nemours & Co., Wilmington, DE, 1986.
[104] S.B. Levy, US Patent 4,510,301, DuPont Co., 1985.
[105] J.G. Drobny, Technology of Fluoropolymers, second ed., CRC Press, Boca Raton, FL, 2009, p. 82.
[106] Aclar® Barrier Films, Allied Signal Plastics, Morristown, NJ.
[107] D.P. Carlson, W. Schmiegel, in: W. Gerhartz (Ed.), Ullmann's Encyclopedia of Industrial Chemistry, vol. A11, VCH Publishers, Weinheim, 1988, p. 413.
[108] A.B. Robertson, Appl. Polym. Symp. 21 (1973) 89.
[109] G. Stanitis, in: J. Scheirs (Ed.), Modern Fluoropolymers, John Wiley & Sons, Chichester, 1997, p. 534.
[110] G. Stanitis, in: J. Scheirs (Ed.), Modern Fluoropolymers, John Wiley & Sons, Chichester, 1997, p. 538.
[111] J.S. Humphrey, J.E. Dohany, C. Ziu, Ultrapure Water, First Annual High Purity Water Conference & Exposition, April 12–15, 1987, p. 136.
[112] W.D. Benzinger, D.N. Robinson, US Patent 4,384,047, 1983.
[113] S. Sternberg, US Patent 4,340,482, 1982.
[114] V.L. Lanza, E.C. Stivers, US Patent 3,269,862, 1966.
[115] U.K. Sopory, US Patent 4,318,881, 1982; U.S. Patent 4,591,700, 1986.
[116] M.D. Heaven, Progr. Rubb. Plast. Tech. 2 (1986) 16.
[117] F.E. Bartell, US Patent 3,582,457, 1971.
[118] L. Eyrund, P. Eyrund, F. Bauer, Adv. Ceram. Mater. 1 (1986) 233.
[119] D.E. Hull, B.V. Johnson, I.P. Rodricks, J.B. Staley, in: J. Scheirs (Ed.), Modern Fluoropolymers, John Wiley & Sons, Chichester, 1997, p. 265.
[120] D.E. Hull, B.V. Johnson, I.P. Rodricks, J.B. Staley, in: J. Scheirs (Ed.), Modern Fluoropolymers, John Wiley & Sons, Chichester, 1997, p. 266.
[121] D.E. Hull, B.V. Johnson, I.P. Rodricks, J.B. Staley, in: J. Scheirs (Ed.), Modern Fluoropolymers, John Wiley & Sons, Chichester, 1997, p. 267.
[122] D.E. Hull, B.V. Johnson, I.P. Rodricks, J.B. Staley, in: J. Scheirs (Ed.), Modern Fluoropolymers, John Wiley & Sons, Chichester, 1997, p. 268.
[123] R.N. Haszeldine, coworkers, Polymer (Guildf) 14 (1973) 122.
[124] D.P. Carlson, W. Schmiegel, in: W. Gerhartz (Ed.), Ullmann's Encyclopedia of Industrial Chemistry, vol. A11, VCH Publishers, Weinheim, 1988, p. 416.
[125] J.L. Hecht, US Patent 3,265,678, E.I. du Pont de Nemours & Co., 1966.
[126] R.E. Uschold, US Patent 5,229,480, E.I. du Pont de Nemours & Co., 1993.

[127] J.G. Drobny, (Chapter 1) Handbook of Thermoplastic Elastomers, second ed., Elsevier, Oxford, 2014.
[128] W. Michel, Kunstst. Ger. Plast. 79 (1979) 964.
[129] P. Resnick, W.H. Buck, in: J. Scheirs (Ed.), Modern Fluoropolymers, John Wiley & Sons, Chichester, 1997, p. 397.
[130] P. Resnick, W.H. Buck, in: J. Scheirs (Ed.), Modern Fluoropolymers, John Wiley & Sons, Chichester, 1997, p. 398.
[131] N. Sugyiama, in: J. Scheirs (Ed.), Modern Fluoropolymers, John Wiley & Sons, Chichester, 1997, p. 542.
[132] Tedlar® Technical Information, Publication H-49725 (6/93), E.I. du Pont de Nemours & Co.
[133] Tedlar® Technical Information, Publication H-4971 (6/93), E.I. du Pont de Nemours & Co.
[134] Tedlar® Technical Information, Publication H-09905 (6/93), E.I. du Pont de Nemours & Co.
[135] M. Tatemoto, US Patent 4,243,770, Daikin, Kogyo Co., 1981.
[136] M. Tatemoto, US Patent 5,198,502, Daikin, Kogyo Co., 1993.
[137] DAI-EL™ Fluoro TPV, Fluoroplastics and Fluoroelastomers in Fusion Expanding Their Performances, Report ERC-4, Daikin Industries, Ltd., 2010. <www.daikin-america.com>, 2013.
[138] CEFRAL SOFT®, Technical Information and Technical Data, Central Glass Co. Ltd. <www.cgc-jp.com/products>.
[139] E.H. Park, US Patent 7,135,527, Freudenberg NOK, General Partnership, 2006.
[140] E.H. Park, Paper Presented at TPE TopCon 2010, Society of Plastics Engineers, Akron, OH, September 12–15, 2010.
[141] P. Resnick, W.H. Buck, in: J. Scheirs (Ed.), Modern Fluoropolymers, John Wiley & Sons, Chichester, 1997, p. 401.
[142] P. Resnick, W.H. Buck, in: J. Scheirs (Ed.), Modern Fluoropolymers, John Wiley & Sons, Chichester, 1997, p. 549.
[143] D. Brasure, S. Ebnesajjad, in: H.F. Mark, J.I. Kroschwitz (Eds.), Encyclopedia of Polymer Science and Engineering, vol. 17, John Wiley & Sons, New York, 1989, p. 480.
[144] L.E. Scoggins, US Patent 3,627,854, 1971.
[145] J.P. Stalings, R.A. Paradis, J. Appl. Polym. Sci. 14 (1970) 461.
[146] L.R. Barton, US Patent 2,953,818, 1960.
[147] K.U. Usmanov, et al., Russ. Chem. Rev. 46 (5) (1977) 462.
[148] G. Guerra, F.E. Karasz, W.J. MacKnight, Macromolecules 19 (1986) 1935.
[149] L.E. Wolinski, US Patent 3,274,088, 1966.
[150] D. Brasure, S. Ebnesajjad, in: H.F. Mark, J.I. Kroschwitz (Eds.), Encyclopedia of Polymer Science and Engineering, vol. 17, John Wiley & Sons, New York, 1989, p. 486.
[151] P. Resnick, W.H. Buck, in: J. Scheirs (Ed.), Modern Fluoropolymers, John Wiley & Sons, Chichester, 1997, p. 413.

[152] P. Resnick, W.H. Buck, in: J. Scheirs (Ed.), Modern Fluoropolymers, John Wiley & Sons, Chichester, 1997, p. 414.
[153] P. Resnick, W.H. Buck, in: J. Scheirs (Ed.), Modern Fluoropolymers, John Wiley & Sons, Chichester, 1997, p. 417.
[154] Public Works III, (1980) 78.
[155] Du Pont Magazine 82 (March–April 1988) 8.
[156] Du Pont Magazine 82 (May–June 1988) 13.
[157] D. Brasure, S. Ebnesajjad, in: H.F. Mark, J.I. Kroschwitz (Eds.), Encyclopedia of Polymer Science and Engineering, vol. 17, John Wiley & Sons, New York, 1989, p. 488.
[158] G.L. Schmutz, Circuits Manuf. 23 (1983) 51.
[159] N. Tatemoto, T. Shimidzu, in: J. Scheirs (Ed.), Modern Fluoropolymers, John Wiley & Sons, Chichester, 1997, p. 563.
[160] N. Tatemoto, M. Tomoda, M. Kawachi, Japanese Patent 62635, Kokai Tokyo Koho Co., 1984.
[161] C. Kawashima, S. Koga, Jpn. Plast. 39 (1988) 38.
[162] T.C. Cheng, B.A. Kaduk, A.K. Mehan, US Patent 4,935,467, Raychem Corp., 1988.
[163] K. Kawamura, C. Kawashima, S. Koga, US Patent 4,749,610, 1988.
[164] N. Tatemoto, T. Shimidzu, in: J. Scheirs (Ed.), Modern Fluoropolymers, John Wiley & Sons, Chichester, 1997, p. 575.
[165] H.G. Floch, P.F. Belleville, SPIE 1758 (1992) 135.
[166] D.E. Keys, US Patent 5,061,024, 1991.
[167] E.N. Squire, US Patent 4,530,569, 1985.
[168] C.-C. Cho, R.M. Wallace, L.A. Files-Sesler, J. Electron. Mater. 23 (1994) 827.
[169] N. Sugyiama, in: J. Scheirs (Ed.), Modern Fluoropolymers, John Wiley & Sons, Chichester, 1997, p. 550.
[170] N. Sugyiama, in: J. Scheirs (Ed.), Modern Fluoropolymers, John Wiley & Sons, Chichester, 1997, p. 553.

# 3 Polytetrafluoroethylene Films

## 3.1 Manufacturing Methods for the Production of Polytetrafluoroethylene Films and Sheets

As pointed out in the previous chapter, polytetrafluoroethylene (PTFE) is manufactured and offered to the market in essentially three forms, namely, as granular resins, as fine powders, and as aqueous dispersions. Although they are all chemically high-molecular-weight PTFE with an extremely high-melt viscosity, each of them requires a different processing technique. This chapter deals with the fabrication methods that are being used for producing films and sheets from granular resins by skiving, producing extruded films from fine powders, and casting of films from aqueous dispersions.

### 3.1.1 Skived Polytetrafluoroethylene Films and Sheets

Skived PTFE films and sheets are produced from granular PTFE in two separate steps: the first one being compression molding of PTFE granular resins using a technique similar to that which is common in powder metallurgy. In this step, large cylinders called *billets* are produced. In the second step, billets are subjected to a process, called *skiving*, which is essentially machining to produce thin sheets and films using a sharp cutting tool.

#### *3.1.1.1 Compression Molding*

The basic molding process for the PTFE consists of the following three important steps: preforming, sintering, and cooling [1]. In the *preforming* step the PTFE molding powder is compressed in a mold at ambient temperature into compacted form, with sufficient mechanical integrity for handling and sintering, called *preform.* The preform is then removed from the mold and *sintered* (heated above the crystalline melting point of the resin). During sintering the resin particles coalesce into a strong homogeneous structure. During the subsequent step, *cooling*, the product hardens while becoming highly crystalline. The degree of crystallinity depends mainly on the rate of cooling.

**Applications of Fluoropolymer Films.** DOI: https://doi.org/10.1016/B978-0-12-816128-9.00003-9

In the process called *preforming*, the loose bed of the molding powder used for the films is compacted in a mold placed in a hydraulic press. The molds used for preforming are similar to those for thermosets or powdered metals and are made from tool steel and are nickel or chrome plated for corrosion protection [2]. The most common shape of a preform is a cylinder, commonly called a *billet.* Presses used for the compression molding must have good controls for a smooth pressure application, be capable of applying specific pressures up to 100 N/mm$^2$ (14,500 psi), have sufficient daylight, and allow easy access to molds. Typically, virgin resins require specific pressures up to 60 N/mm$^2$ (8700 psi) and filled compounds up to 100 N/mm$^2$ (14,500 psi). A press for large preforms is shown in Fig. 3.1.

**Figure 3.1** Compression-molding press for PTFE billets. *PTFE*, Polytetrafluoroethylene.

PTFE resins exhibit a first-order transition at 19°C (66°F) due to a change of crystalline structure from triclinic to hexagonal unit cell (see Section 2.1.3). A volume change of approximately 1% is associated with this transition (Fig. 3.2). Another consequence of this transition is that the resin has a better powder flow below 19°C but responds more poorly to the preform pressure. Billets prepared below this transition are weaker and tend to crack during sintering. For this reason the resin should be conditioned at about 21°C–25°C (70°F–77°F) overnight before preforming to prevent that. The preforming operation should be done at room temperature, preferably higher than 21°C (77°F). Preforming at higher temperatures is sometimes useful to overcome press-capacity limitations. As the temperature is raised, the resin particles exhibit higher plastic flow and consequently can be more easily compacted and become more responsive to preform pressure. Mold filling is another key factor in the quality of the final product. It has to be uniform which is achieved by breaking up lumps of resin with a scoop or screening. The full amount of the powder has to be charged into the mold before the pressure is applied; otherwise, contamination, layering, or cracking at the interfaces may occur on sintering.

During the compression of the PTFE powder, both plastic and elastic deformations occur. At low pressures the particles slip, slide, and tumble in place to align themselves to the best possible array for packing. With increasing pressure, contact points between adjacent particles are established and further enlarged by plastic deformation. Plastic deformation also eliminates internal particle voids.

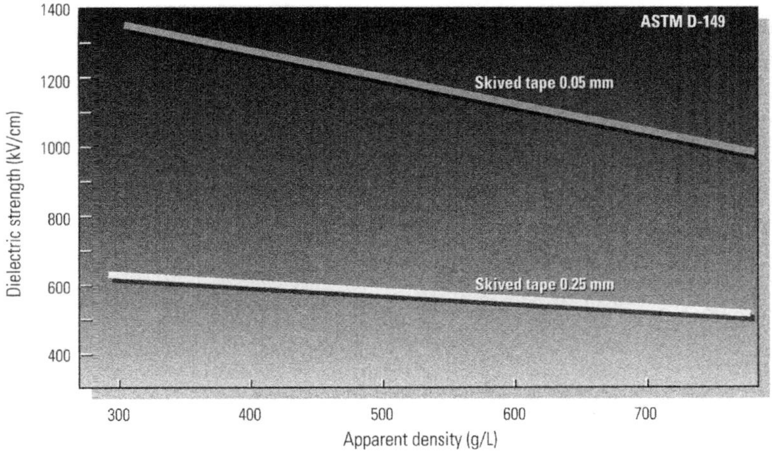


**Figure 3.2** Transition point and linear thermal expansion of PTFE. *PTFE*, Polytetrafluoroethylene. Courtesy DuPont Co.

When the maximum pressure required for compression is reached, it is held for a certain time, which is referred to as *dwell time*. This time is required for the transmittal of pressure throughout the preform and for the removal of entrapped air. Too short a dwell time can cause density gradients in the preform, which may lead to "hour-glassing" and property variation in the sintered billet. Incomplete removal of trapped air can cause microfissures and/or worsened properties. Dwell time is dependent on the rate of pressure application and on the shape and mass of the billet. Generally, 2–5 min/10 mm or 0.5 in. finished height for small and 1–1.5 min/10 mm or 0.5 in. finished height for large billets give satisfactory results [3]. Pressure release after the dwell period should be very slow until the initial expansion and relaxation has taken place. Typically, this is done with a bleeder valve or a capillary. Sudden pressure decay can result in microcracks or visible cracks as the still-entrapped air expands.

After preforming, there are still residual stresses and entrapped air in the preformed part, which invariably causes cracking mainly during the initial stage of sintering. The removal of entrapped air or "degassing" requires some time called *resting time*, which depends mainly on the wall thickness.

The next step in the fabrication of most PTFE films is *sintering*. The purpose of the sintering operation is to convert the preform into a product with increased strength and reduced fraction of voids. In this technology, massive billets are generally sintered in an air-circulating oven (Fig. 3.3) heated to 365°C–380°C (689°F–716°F). Both the sintering temperature and time have a critical effect on the degree of coalescence, which in turn affects the final properties of the product.

In this case, sintering has two stages and consists of a variety of processes. During the first stage, the preform expands up to about 25% of its volume as its temperature is increased to and above the melting point of the virgin resin (about 342°C or 648°F). The next stage is coalescence of particles in which voids are eliminated. After that the contacting surfaces of adjacent particles fuse and eventually melt. The latter process gives the part its strength. After reaching the melting point the resin changes from a highly crystalline material to an almost transparent amorphous gel. When the first stage is completed, the billet becomes translucent, but it will require additional time to become fully sintered. This time depends not only on the sintering temperature but also on conditions at which the part was preformed and on the type of resin. High pressures and small particle size facilitate fusion. As it is with

**Figure 3.3** Typical production sintering oven.

preform pressure, sintering time eventually reaches a point beyond which there is no significant improvement in physical properties to justify longer sintering [4].

During sintering, some degradation of the polymer takes place. Prolonging the process beyond the required time or using very high sintering temperature will invariably result in excessive degradation and considerable worsening of properties. To achieve a uniform heat distribution in the oven, turbulent airflow is required. Variability of heat distribution can cause billet distortion or even cracking. Massive billets should be loaded into the oven at maximum 100°C (182°F) to avoid thermal shock. A holding time of 1–2 hours before heat-up is usually required for the temperature to reach equilibrium.

The heating rate is very critical for the quality of the final product. Because of the very low thermal conductivity of the PTFE resin, the billets have to be heated slowly to the sintering temperature or cracking may occur even before the resin is fully melted. The highest heating rate a given preform will tolerate depends on a complex interaction of many factors. Major parameters include thermal gradient (difference between the ambient temperature in the oven and the temperature in the midpoint of the preform wall) and the rate of internal stress relaxation. The thermal gradient, in turn, is related to the heating rate and the wall thickness. Internal stresses in the preform originate in the preforming process and are dependent mainly on the preform pressure, closure rate, and preform temperature. Normal heating rate for large billets is 28°C (50°F)/h up to 300°C (572°F) at which point it is reduced to 6°C–10°C (11°F–18°F)/h.

A proper rate for a given set of billet geometry and preforming conditions is usually determined experimentally. Another way to minimize thermal gradient is to introduce a series of hold periods. In this method a higher heating rate is used in the early phase of the heat-up cycle. Hold periods used between temperatures of 290°C and 350°C (554°F and 662°F) ensure a minimum temperature gradient through the melting transition and minimize any tendency of cracking due to about 10% volume change associated with melting.

At the point when the billet is in the gel state, the particles coalesce and the voids are eliminated. However, the rate of sintering near the melting point is very slow. To achieve more commercially acceptable sintering rates, the temperatures used for that purpose are in the range between 365°C and 380°C (689°F and 716°F). For massive billets, temperatures above 385°C (725°F), for virgin PTFE, and 370°C (698°F), for filled compounds, should be avoided since thermal degradation above these temperatures becomes significant. The time at the peak temperature depends on the wall thickness, type of resin used, and the method of sintering. Generally, prolonged sintering times have beneficial effects on properties, particularly on dielectric strength, provided that no significant degradation takes place. Typical times for a complete sintering process are fairly constant, about 2 hours after the resin has reached its optimum sintering temperature. Once the oven has reached the sintering temperature, it takes about 1–1.5 hours to transmit the heat through each centimeter of thickness. As a rule of thumb, the time at sintering temperature should be 1.0 h/cm or 0.4 in. of diameter for solid billets and 1.4 h/cm or 0.4 in. of wall thickness for billets with a small hole in the middle.

After sintering, the molten billet is cooled to the room temperature in a controlled fashion. As the freezing range of 320°C–325°C (608°F–617°F) is reached, crystallization starts. The degree of crystallization in the cooled-down part depends on the cooling rate. Since a majority of properties depends on the degree of crystallinity, the cooling rate has to be closely controlled to achieve the desired results [3]. The effect of cooling rate on crystallinity is shown in Table 3.1. Melt strength and wall thickness are the key factors in determining the cooling rate. Typically, cooling rates between 8°C (14°F)/h and 15°C (27°F)/h are satisfactory for larger billets. Because of low thermal conductivity, slow cooling rates are necessary to avoid cracking due to excessive thermal

**Table 3.1 Effect of Cooling Rate on Polytetrafluoroethylene Crystallinity.**

| Cooling Rate (°C/min) | % Crystallinity |
|---|---|
| Quenched in ice water | 45 |
| 5 | 54 |
| 1 | 56 |
| 0.5 | 58 |
| 0.1 | 62 |

*Source*: Compression Moulding, Technical Information, Publication H-59487, 05/95, DuPont, Wilmington, DE, 1995.

gradients. This is particularly important during the transition in the freezing zone since stresses, caused by a rapid volume change, can tear the melt apart. Therefore slower rates are maintained until the inside of the wall is below the freezing point and the center of the billet is crystallized. Then, faster cooling rates, about 50°C (90°F)/h can be used since the sintered part can tolerate higher thermal gradients [3].

Massive moldings often require *annealing* during the cooling period to minimize thermal gradients and to relieve any residual stresses. The temperature range for annealing is typically from 290°C to 325°C (554°F–617°F). The temperature at which the annealing is carried out is very critical. If annealing is done in the temperature range 310°C–325°C (590°F–617°F), the molding exhibits a high degree of crystallinity, which may not always be desirable. The product is highly opaque and has low tensile strength, high stiffness, and high specific gravity. If a lower degree of crystallinity is desired, annealing should be done at 290°C (554°F) [3]. An example of finished billets is in Fig. 3.4.

#### 3.1.1.2 Skiving Process

Films and sheets are produced by *skiving*, which is "peeling" of the billet in a similar fashion as a wood veneer. A grooved mandrel is pressed into a billet and the assembly is mounted onto a lathe. A sharp cutting tool is used to skive a continuous tape of a constant thickness, the arrangement of which is shown in Fig. 3.5. The range of thickness of films and sheets produced by skiving is typically from about 25 μm to 3 mm (0.001–0.125 in.) [5]. A modern, high-performance skiving machine, capable of machining billets up to 1500 mm (60 in.) wide, is shown in Fig. 3.6.

**Figure 3.4** Medium-sized sintered billets.

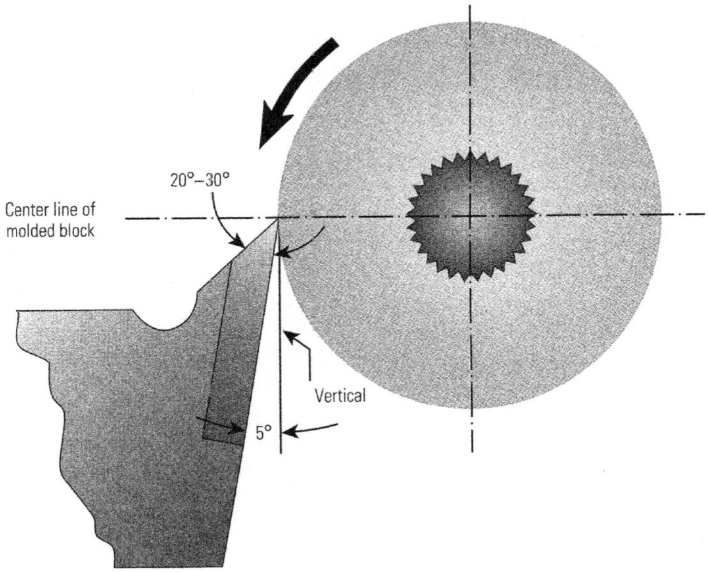


**Figure 3.5** Typical arrangement of skiving knife. Courtesy DuPont Co.

**Figure 3.6** Modern skiving machine. Courtesy Dalau, Inc.

### 3.1.2 Extruded Unsintered Polytetrafluoroethylene Films and Tapes

Unsintered films and tapes are produced from PTFE fine powders. Fine powder resins are extremely sensitive to shear, and the sheared polymer cannot be processed. Because of this, they have to be handled with great care during transport and processing. Most commonly, fine powder resins are processed in the form of a "paste." Such a paste is prepared by mixing the powder with 15%–25% hydrocarbon *lubricant*, such as kerosene, white oil, or naphtha, with the resultant blend appearing much like the powder alone [6].

Fine powder resins are shipped in specially constructed drums that typically hold 23 kg (50 lb) of resin. These shallow, cylindrical drums are designed to minimize compaction and shearing of the resin during shipment and storage. To assure further that the compaction is kept at an absolute minimum; the resin must be kept at a temperature below 19°C (66°F), its transition point during shipping and warehouse storage. Prior to blending with lubricants, the resin should be stored below its transition temperature for 24 hours. A safe storage temperature for most resins is 15°C (60°F). Generally, the particles form agglomerates, spherical in shape with an average size of 500 μm. If lumps have formed during shipping, the resin should be poured through a four-mesh screen immediately prior to blending. To prevent shearing the screen should be vibrated gently up and down. Sharp objects, such as scoops, should not be used to remove the resin because they could shear and ruin the soft resin particles. It is best to avoid screening unlubricated powder unless it is absolutely necessary.

Lubricants enable PTFE fine powders to be processed on commercial equipment. Liquids with a viscosity between 0.5 and 5 cP (0.5 and 5 mPa s) are preferred, although more viscous liquids are used occasionally. When selecting a lubricant, its ability to be incorporated easily into the blend and to vaporize completely and rapidly in a later processing step without leaving residues, which would discolor the product or adversely affect its properties, is important. The amount of lubricant added is typically 16%–19% of the total weight of the mix.

For colored products, pigments may be added during blending. These can be added dry directly to the powder prior to the addition of the lubricant, or as a wet blend in the lubricant. In the latter case the pigment dispersion is added with the remaining lubricant.

All blending operations must be performed in an area in which the temperature is maintained below the PTFE transition temperature of 19°C (66°F) and the relative humidity should be kept at approximately 50%. A high level of cleanliness and an explosion-proof environment are additional requirements to assure high quality and safety.

Commercial batches, of sizes typically 10–136 kg (22–300 lb), are often prepared in twin-shell blenders (Fig. 3.7) by tumbling for 15 minutes at 24 rpm. The blend is screened again, transferred into a storage vessel, and allowed to age for at least 12 hours [7].

The properly aged lubricated powder is usually preformed at room temperature into a billet of the size required by the equipment in which it is to be later processed. Preforming removes air from the material and compacts is so that it has a sufficient integrity for handling during the manufacturing process. Preforming pressures are on the order of 0.7 MPa (100 psig) in the initial stage of the cycle and may increase up to 2 MPa (300 psig). Higher pressures do not increase compaction and may cause the lubricant to be squeezed out [7]. The compaction rate is initially up to 250 mm (10 in.)/min and is reduced toward its end. The finished preform is rather fragile and must be stored in a PVC or PMMA tube for protection against damage and contamination.

#### 3.1.2.1 Extrusion

The extrusion step is performed at temperatures above 19°C (66°F), the first transition point of the resin, where it is highly deformable and

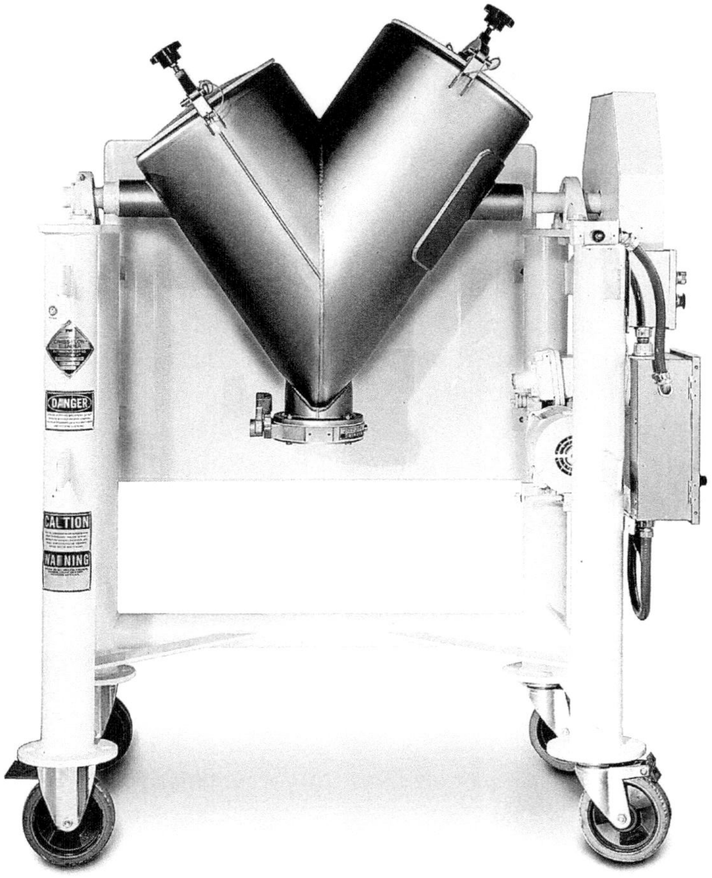


**Figure 3.7** Twin-shell blender. Courtesy Patterson Kelley Co.

can be extruded smoothly. The resin preform is placed in the extrusion cylinder, which is kept at a temperature of 38°C (100°F) for several minutes to heat up to the higher temperature.

Unsintered films and tapes, for thread sealing and cable wrapping, and unsintered rods, for packing together, represent one of the largest applications for this technology. Other applications are sintered thin-walled tubing and wire coating.

Extrusion of an unsintered rod is the simplest process for paste extrusion of PTFE fine powders. The extrusion is done by a simple hydraulic ram extruder with a total available thrust ranging from 10 to 20 tons. The ram speed is adjustable up to a maximum of 50–100 mm/min (2–4 in./min). The ram forces the lubricated powder through the orifice of the die. The head of the ram is usually fitted with a PTFE seal to prevent the polymer from flowing back along the ram. The surfaces of the extrusion cylinder and die are made from corrosion-resistant steel and are highly polished. The pressures during the rod extrusion are normally in the range of 10–15 MN/m$^2$ (1500–2000 psi). The diameter of the extrusion cylinder is designed to accommodate the required size of the extrudate and the type of polymer used. The usual extrusion cylinder diameters are between 40 and 150 mm (1.5 and 6 in.). The die consists of two parts, a conical and a parallel section. The conical part is more important, as the reduction in area between cone entry and exit determines the amount of work done on the polymer and hence, to a high degree, the properties of the extrudate. The cone angle has also some effect on the extrudate and is most commonly 30 degrees, although larger angles are used for large-diameter cylinders. The length/diameter ratio of the die parallel (die land), which also has some effect on the properties of the extrudate, may vary from 5 to 10 times the exit diameter. An example of conditions for rod extrusion is in Table 3.2.

**Table 3.2 Example of Extrusion Conditions for a Rod (Diameter 10 mm).**

| | | |
|---|---|---|
| **Die Tube** | | |
| Diameter | | 10.6 mm |
| Unheated length at the top of die tube | | 90 mm |
| Heated length | | 900 mm |
| Unheated length at the bottom of die tube | | 400 mm |
| Total length | | 1550 mm |
| Heated length/diameter | | 85:1 |
| Water cooling | | Over top 60 mm |
| *Heating arrangements* | | Four separately controlled heated zones, each with 2/1.5 kW heater bands. |
| Temperature profile (top) | Zone 1 | 380°C |
| | Zone 2 | 400°C |
| | Zone 3 | 400°C |
| (Bottom) | Zone 4 | 350°C |

*Source*: Adapted from J.G. Drobny, Technology of Fluoropolymers, second ed., CRC Press, Boca Raton, FL, 2009, p. 71 [8].

**Table 3.3** **Typical Properties of Skived Films.**

| Property | Value |
|---|---|
| Thickness, mil (μm) | 0.5–125 (12.5–3125) |
| Specific gravity | 2.14–2.19 |
| Tensile strength, minimum, psi (MPa) | 3500 (24.1) |
| Elongation, minimum (%) | 300 |
| Dielectric strength (V/mil) | 1680 |
| Hardness, Shore D | 54 |

*Source*: <www.films.saint-gobain.com>, 2017.

**Table 3.4** **Typical Properties of Standard Cast Films [10].**

| Property | Value |
|---|---|
| Density (g/cm$^3$) | 2.14–2.19 |
| Tensile strength, psi (MPa) | 3500 (24.1) |
| Elongation at break (%) | 300 |
| Dielectric strength (V/mil) | 1680 |
| Hardness, Shore D | 54 |

**Table 3.5** **Comparison of Cast- and Skived-Film Properties [10].**

| | | Tensile Strength (MPa) | | Elongation at Break (%) | | Elastic Modulus (MPa) | |
|---|---|---|---|---|---|---|---|
| **Film Type** | **Thickness (μm)** | **MD** | **TD** | **MD** | **TD** | **MD** | **TD** |
| Skived | 76 | 52.3 | 40.4 | 450 | 360 | 469 | 517 |
| Cast | 68 | 36 | 34 | 530 | 510 | 434 | 434 |

The extruded rod is converted into tape by calendering. Rod for packing is sometimes used with the lubricant still in it. If the lubricant is to be removed, this may be done in a simple in-line oven immediately after the extruder, or in a separate batch oven [9] (Tables 3.3–3.5).

#### 3.1.2.2 Manufacture of Unsintered Tape

An example of the application of a thin unsintered film is the unsintered tape. The process of unsintered tape, involving the conversion of the extruded rod, consists of the following sequence of operations [8]:

1. Calendering
2. Removal of lubricant
3. Slitting
4. Reeling

The rod used in this method has a relatively large diameter, typically 10–15 mm (0.4–0.6 in.) and is calendered in a single pass to a tape about 100–200 mm (4–8 in.) wide and 0.075–0.01 mm (0.003–0.004 in.) thick, which is subsequently slit to several tapes of desired width. No advantage has been found in using multiple-stage calendering [8].

The rod is fed into the nip of the calender by means of a guide tube that prevents it from wandering and causing consequent variation in the width of the calendered tape. A tape with straight edges and controlled width is produced when a fishtail guide made from metals (e.g., aluminum) or plastics (e.g., acetal) is used successfully [8].

The calender rolls are heated to temperatures up to 80°C (176°F) to produce a smooth, strong tape. They may be heated either electrically or by circulating hot water or oil [8]. Production speeds of 0.05–0.5 m/s (10–100 ft./min) are quite common.

The thickness of the calendered tape depends mainly on the calender nip setting. The tape width depends on several factors, such as lubricant type and content, reduction ratio and die geometry during the extrusion, as well as the speed and temperature of the calender rolls [8].

Subsequently, the lubricant is removed by passing the calendered tape through a heated tunnel oven. If the lubricant used in the mixture has a very high boiling point, such as mineral oil, it may be removed by passing the calendered tape through a degreasing bath containing hot vapor of trichloroethylene or other suitable solvents. The slow output rate of this process and safety and health hazards associated with this method rarely offset the advantages of using heavy oils, such as virtually no loss of lubricant by evaporation between extrusion and calendering and the amount of work done on the polymer as a result of the much higher viscosity of the paste. Therefore this technique is seldom used, only in cases where it is absolutely necessary.

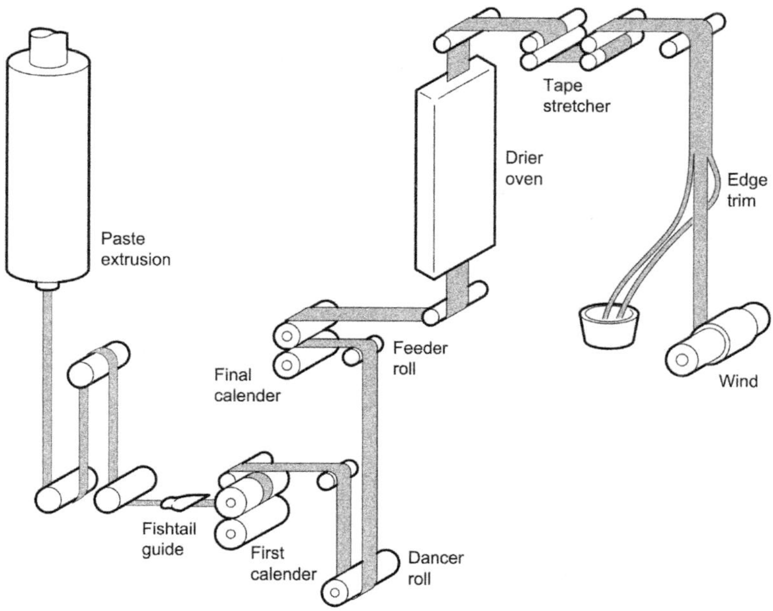


**Figure 3.8** Schematic of process for production of an unsintered tape. Courtesy DuPont Co.

After the lubricant is removed, the tape is slit to the desired width by leading it under a slight tension over stationary or rotating cutting blades. The slit tape is reeled onto small spools [8]. An outline of the process is shown in Fig. 3.8.

## 3.1.3 Process for Unsupported Cast Polytetrafluoroethylene Films

As pointed out earlier, PTFE homopolymer cannot be processed by melt extrusion because of its extremely high-melt viscosity. Thus other methods, such as skiving from compression-molded and sintered billets (see Section 3.1.1), extrusion of unsintered films from fine powders (Section 3.1.2), and by casting from aqueous dispersions, were developed to prepare films. The original method for casting unsupported films from PTFE dispersions [11] employs polished stainless-steel belts, which are dipped into a properly compounded dispersion. The thin coating of the liquid is then dried and the dry powdery layer is subjected to baking and sintering. To obtain a good-quality film, the thickness of the film has to be below the critical value to prevent a condition called *mudcracking*.

The equipment used is a vertical coater with heated zones [11]. The speed of the belt is slow, about 0.3–1 m/min (1–3 ft./min) and there are no applicators used to remove excess dispersion [12]. The amount of coating picked up by the belt is controlled mainly by the solid content of the dispersion and by the belt speed. Immediately after the dipping, the coated belt is heated in the first (drying) zone at approximately 120°C (248°F) to evaporate the water and then dry the resin. In the next (baking) zone the belt and resin are heated to a temperature about 327°C (620°F) to remove residual wetting and thickening agents. The third (sintering) zone is heated to a temperature above 350°C (662°F), and in that zone, the resin is sintered into a continuous film. By repeating this sequence of deposition, drying, baking, and sintering, the film thickness can be built up to 0.003 in. (0.076 mm) or more. The film is then stripped from the metal base. A production machine is built with multiple stages. Thus after a film is sintered, it is recoated in the next stage. At the end of the machine, there is a device designed to strip the finished films from both sides of the belt and wind them up into rolls.

The advantage of this method is that each layer can be made from a different type of dispersion. For example, clear and pigmented layers can be made or the top layer can be prepared from a FEP or PFA dispersion to obtain films that can be heat-sealed or laminated. In fact, films with both surfaces that are heat-sealable can be produced by this method. In such an instance, PFA is applied as the first coat onto the belt and FEP as the last coat, because PFA can be stripped from the steel belt, whereas FEP would adhere to it and is impossible to strip the finished film [13]. Another possibility is to make films with an unsintered last coat, which can be used for lamination with substrates coated with unsintered PTFE using the lamination method described in Section 3.1.6.1. The schematic of this coater and all its possible embodiments are shown in Fig. 3.9. These embodiments, covered by US Patent 2.852,811 [11], are single-stage coater (Fig. 1), multistage coater (Fig. 2), coater with endless belt (Fig. 3), and the multistage coater including heating setups for drying, baking, and sintering the thermoplastic films.

An improved process and equipment for cast PTFE films have been developed, which have considerably higher productivity than the method and equipment described earlier [13]. The process essentially uses a vertical coater with multiple stages. The carrier belt has to be made from a material of low thermal mass, which can tolerate repeated

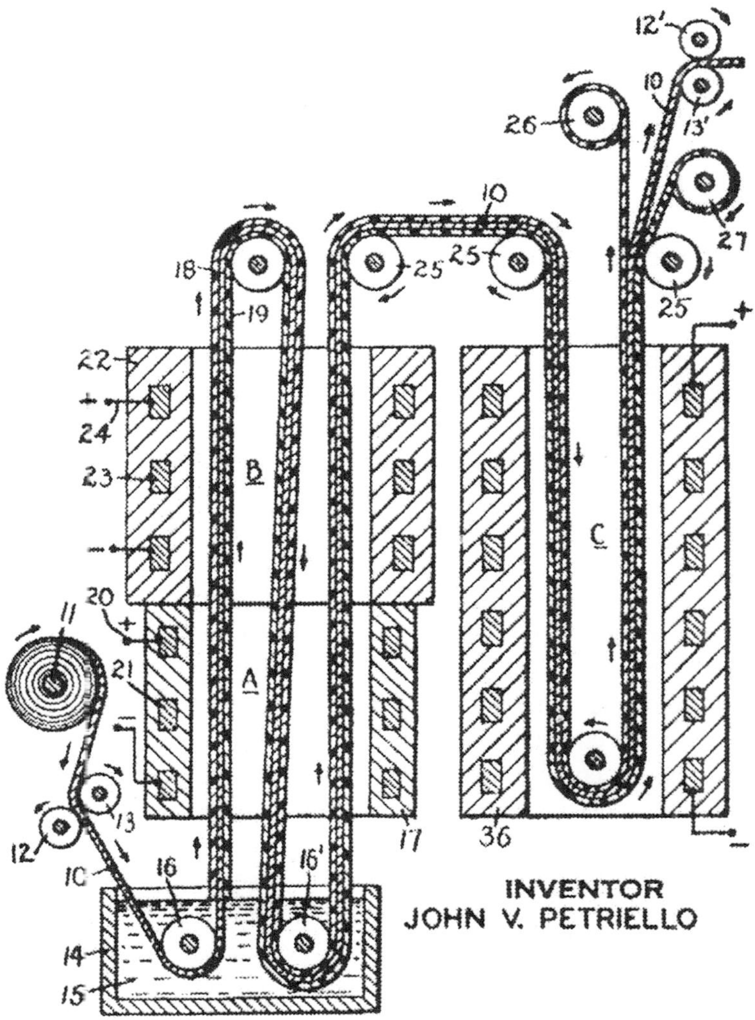


**Figure 3.9** Schematic of the steel belt coater for the production of PTFE films [11]. *PTFE*, Polytetrafluoroethylene.

exposure to the sintering temperature and has surface properties such that it can be wetted by the dispersion, yet the film can be stripped without being stretched. There are several possible belt materials, but KAPTON HN (DuPont) film pretreated by a fluoropolymer was found to be particularly suitable because of its heat resistance, dimensional stability, and surface characteristics [14]. The production speed used in this process is 3–10 m/min (9–30 ft./min).

Unlike in the coater with steel belts, in this process, equipment applicators, such as metering wire-wound bars (5), are used in the metering zone (4) to remove excess dispersion. These may be designed to rotate to assure a better, more uniform coating. The dip tank (3) is similar to that used in fabric-coating machines, which contains an immersed rotary dip bar. To prevent coating defects due to shear, the dip tanks have double walls and are chilled by circulated chilled water to temperatures below 19°C (66°F), the first-order transition temperature of PTFE. This coater (100) has two heating zones, one for drying (6) and the other for baking/sintering (7), and one cooling plenum (8) from which it can be directed to the dip tank of the next stage or to the stripping apparatus. The temperature settings for the drying zone are typically 250°F (121°C) and for the baking/sintering zone 680°F–710°F (360°C–377°C) in order to accommodate the faster operating speed. The heating is a combination of conventional heat source (9) and a radiant heat source (10). The schematic of this coater is illustrated by Fig. 3.10.

The dispersions used in this process are formulated in such a way that they wet the carrier sufficiently and tolerate higher shear at the wire-wound bars due to a relatively high coating-belt speed. This is accomplished by a combination of nonionic surfactants (e.g., octylphenoxy polyethoxy ethanol) and fluorosurfactants. The original PTFE dispersions can be colored by adding heat-resistant pigments dispersed with appropriate dispersion agents. Additional modification, if necessary, is an increase of viscosity with the use of special viscosifiers. This equipment and corresponding process are discussed in detail in Refs. [13,14].

It should be noted that that the coaters discussed here can be used to process not only PTFE aqueous dispersions but also aqueous dispersions of FEP, PFA/MFA [15] as well as those of PVDF and THV and their combinations. The dispersion of polymers with lower melting temperatures can be cast on other coating belts, such as polyethylene terephthalate [16], which tolerate their processing temperatures.

### 3.1.4 Process for Supported Cast Polytetrafluoroethylene Films

This process involves, essentially, dipping of fabrics or other substrates into aqueous fluoropolymer dispersions and building up a film of required thickness by repeating the dipping cycle [10]. In this technology the

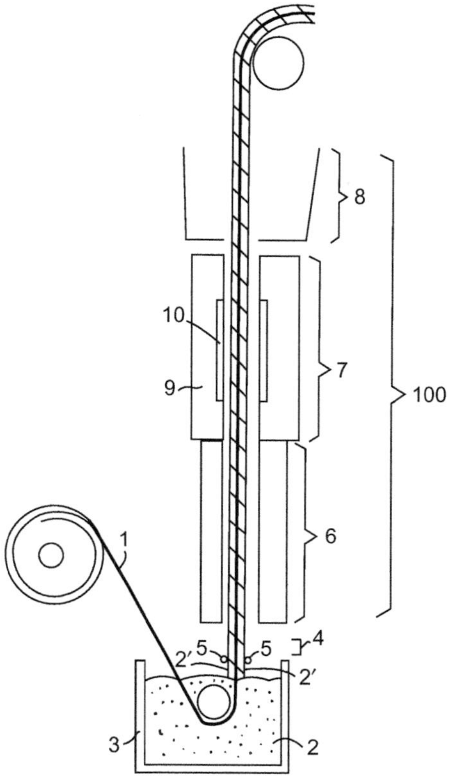


**Figure 3.10** Schematic of the high-speed coater for fluoropolymer films [14]. The coating tower (100) consists of a metering zone (4), metering bars (5), drying zone (6), bake/fuse zone (7), and cooling plenum.

processing equipment is simpler than that used in the production of unsupported films. The thickness of the film depends on the type of the tool for the removal of excess liquid, which may be a wire-wound bar, smooth bar, or a doctor blade. The fabrics used in this technology, such as glass fabrics and aramid fabrics (Kevlar, Nomex—both DuPont), have to tolerate the high processing temperatures involved [10]. Thin metal foils must be first primed to assure sufficient bond to the first PTFE coat. The coater, frequently referred to as "coating tower," is shown in Fig. 3.11.

## 3.1.5 Laminates

### 3.1.5.1 Laminates Based on Cast Films

The utility of PTFE films is enhanced by their lamination with other polymers and materials. The basic laminating methods include compression in a heated press or continuous compression between two heated rolls.

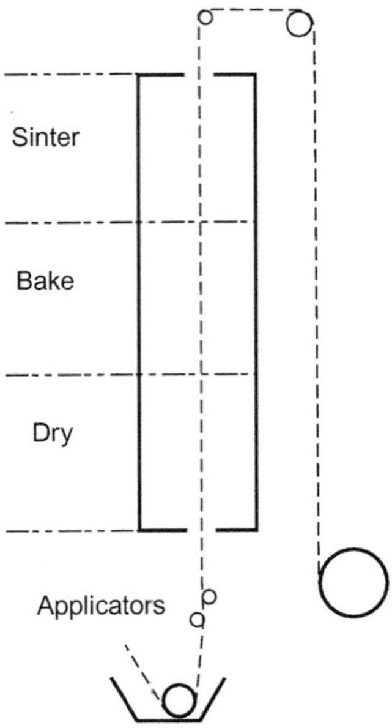


**Figure 3.11** Coating tower for supported fluoropolymer films. Courtesy DuPont Co.

The simplest laminates based on PTFE films are prepared by subsequent casting using methods described in Section 3.1.3, where the film consists of a coating of PFA on one surface and a coating of FEP on the other. Such composite films can be further used for lamination with coated fabrics that have a coating of PFA or FEP on their surface in a heated press or two heated rollers or an equipment design for continuous lamination, which is often of proprietary design.

Another lamination method is based on the use of unsintered (unfused) PTFE applied to the surface of sintered PTFE [17]. As an example, such an unsintered PTFE coating is applied to one side of a sintered PTFE cast film and in the same way, an unsintered coating is applied to a fully sintered PTFE coated fiberglass fabric as discussed in Sections 3.1.3 and 3.1.4, respectively. Both the film and the coated fabric are fed into the nip between two moderately heated rolls (one steel roll—404, the other a packed cotton/wool roll—405) such a way that the two surfaces coated with the unsintered PTFE coating are facing each other. Due to the pressure and shear in the nip, the two unsintered layers fibrillate and form an initial mechanical bond, which is sufficiently strong to resist separation during the remainder of the process. This laminate is subsequently heated to sintering temperature without pressure in the sintering oven (407)

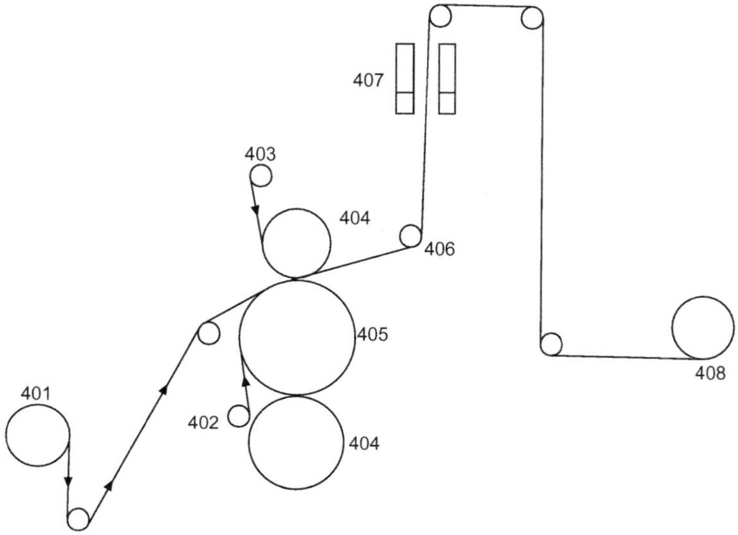


**Figure 3.12** Schematic of the lamination process using unsintered PTFE coating [17]. The equipment consists of pay-off rolls (401, 402, and 403), heated steel roll (404), packed roll (405), sintering oven (407), and take-up roll (408). *PTFE*, Polytetrafluoroethylene.

whereby a strong bond between the two components of the laminate is established. The process is illustrated by Fig. 3.12.

Essentially, the bonding unsintered coating layer can contain certain amounts of other fluoroplastics such as PFA, FEP and/or mineral or carbonaceous fillers and pigments. It allows lamination of PTFE films with coated textiles, certain films, and certain metals. The details regarding this technology are in Ref. [17].

## References

[1] J.G. Drobny, Technology of Fluoropolymers, second ed., CRC Press, Boca Raton, FL, 2009, p. 58.

[2] Compression Moulding, Technical Information, Publication H-59487, 05/95, E. I. Du Pont de Nemours & Co, Wilmington, DE, 1995, p. 3.

[3] Compression Moulding, Technical Information, Publication H-59487, 05/95, E. I. Du Pont de Nemours & Co, Wilmington, DE, 1995, p. 9.

[4] J.G. Drobny, Technology of Fluoropolymers, second ed., CRC Press, Boca Raton, FL, 2009, p. 61.

[5] J.G. Drobny, Technology of Fluoropolymers, second ed., CRC Press, Boca Raton, FL, 2009, p. 65.

[6] T.A. Blanchet, in: O. Olabisi (Ed.), Handbook of Thermoplastics, Marcel Dekker, New York, 1997, p. 990.
[7] Processing Guide for Fine Powder Resins, Publication H-21211-2, E. I. Du Pont de Nemours & Co, Wilmington, DE, 1994, p. 4.
[8] J.G. Drobny, Technology of Fluoropolymers, second ed., CRC Press, Boca Raton, FL, 2009, p. 71.
[9] J.G. Drobny, Technology of Fluoropolymers, second ed., CRC Press, Boca Raton, FL, 2009, p. 70.
[10] S. Ebnesajjad, Fluoroplastics, second ed., Non-Melt Processible Fluoropolymers, vol. 1, Elsevier, Oxford, 2014.
[11] J.V. Petriello, US Patent 2,852,811, September 1958.
[12] J.G. Drobny, Technology of Fluoropolymers, second ed., CRC Press, Boca Raton, FL, 2009, p. 143.
[13] J.G. Drobny, Technology of Fluoropolymers, second ed., CRC Press, Boca Raton, FL, 2009, p. 146.
[14] J.A. Effenberger, K.G. Koerber, M.N. Latorra, J.V. Petriello, US Patent 5,075,065 to Chemical Fabrics Corporation, December 1991.
[15] J.G. Drobny, Technology of Fluoropolymers, second ed., CRC Press, Boca Raton, FL, 2009, p. 147.
[16] J.G. Drobny, Technology of Fluoropolymers, second ed., CRC Press, Boca Raton, FL, 2009, p. 148.
[17] J.A. Effenberger, et al. US Patent 5,141,800, to Chemical Fabrics Corporation, August 1992.

# 4 Films from Melt-Processible Fluoropolymers

## 4.1 Production of FEP Films

### 4.1.1 Materials

FEP resins are available as translucent 2.5 mm (0.1 mm) pellets in the following grades: low melt viscosity, intermediate viscosity, high melt viscosity, extrusion grade, and as aqueous dispersion [1] containing 55 wt.% of hydrophobic, negatively charged FEP particles and approximately 6 wt.% (based on FEP) of a mixture of nonionic anionic surfactants.

### 4.1.2 Equipment

FEP films are produced mainly by melt extrusion on the conventional melt-processing equipment with the modification as mentioned in Section 4.1.3. It should be noted that at the processing temperatures, highly corrosive products are generated. The equipment for the processing of FEP aqueous dispersions is essentially the same as that used for PTFE dispersions.

### 4.1.3 Process Conditions

Processing temperatures used for the extrusion of FEP resins range usually from 315°C to 400°C (600°F to 752°F), [2] at which temperatures highly corrosive products are generated. Therefore the parts of the processing equipment that are in contact with the melt (screw and barrel components) must be made of special corrosion-resistant alloys to assure a trouble-free operation. Also, as with any fluoropolymer, it is necessary to prevent long residence times in equipment and to purge the equipment after the process is finished. In addition, appropriate ventilation during the process is necessary.

**Applications of Fluoropolymer Films.** DOI: https://doi.org/10.1016/B978-0-12-816128-9.00004-0

## 4.2 Production of PFA and MFA Films

### 4.2.1 Materials

PFA resins are available as clear 2.5 mm (0.1 mm) pellets in the following grades: low melt viscosity, intermediate viscosity, high melt viscosity, extrusion grade, and as aqueous dispersion containing 55 wt.% of polymeric particles and approximately 3.5–6 wt.% (based on polymer) of nonionic and anionic surfactants, depending on supplier.

### 4.2.2 Equipment

PFA and MFA films are produced mainly by melt extrusion on the conventional melt-processing equipment, but the parts of the processing equipment that are in contact with the melt must be made of special corrosion-resistant alloys to assure a trouble-free operation. The equipment for the processing of PFA and MFA aqueous dispersions is essentially the same as that used for PTFE dispersions involving coating towers with controlled heating.

### 4.2.3 Process Conditions

Perfluoroalkoxy resins (PFA and MFA) can be processed by standard techniques used for thermoplastics, at temperatures up to 425°C (797°F). High processing temperatures are required, because they exhibit high melt viscosity values with activation energy lower than most thermoplastics, 50 kJ/mol[2]. Extrusion and injection molding are done at temperatures typically above 390°C (734°F) [2] and relatively high shear rates. For these processing methods, PFA grades with high melt flow indexes, that is, with lower molecular weights, are used. Although PFA is thermally a very stable polymer, it still is subject to thermal degradation at processing temperatures, the extent of which depends on temperature, residence time, and the shear rate. Thermal degradation occurs mainly from the end groups; chain scission becomes evident at temperatures above 400°C (752°F) [2], depending on the shear rate. PFA can be extruded into films, tubing, rods, and foams. As with any fluoropolymer, it is necessary to prevent long residence times in equipment and to purge the equipment after the process is finished. In addition, appropriate ventilation is necessary.

## 4.3 Production of ETFE Films

### 4.3.1 Materials

ETFE resins are available as translucent 2.5 mm (0.1 mm) pellets in the following grades: low melt viscosity, intermediate viscosity, high melt viscosity, and extrusion and molding grades.

### 4.3.2 Equipment

At the processing temperatures, highly corrosive products are generated. Therefore the parts of the processing equipment that are in contact with the melt must be made of special corrosion-resistant alloys to assure a trouble-free operation.

### 4.3.3 Process Conditions

ETFE copolymers can be readily fabricated by a variety of melt-processing techniques [3]. They have a wide processing window, in the range of 280°C–340°C (536°F–644°F) and can be extruded into extruded and cast films. Also, as with any fluoropolymer, it is necessary to prevent long residence times in equipment and to purge the equipment after the process is finished. In addition, appropriate ventilation is necessary.

## 4.4 Production of PVDF Films

### 4.4.1 Materials

PVDF resins for melt processing are supplied as powders or pellets with a rather wide range of melt viscosities. Lower viscosity grades are used for injection molding of complex parts, while the low-viscosity grades have high enough melt strength for the extrusion of profiles, rods, tubing, pipe, film, wire insulation, and monofilament. PVDF extrudes very well, and there is no need to use lubricants or heat stabilizers [4].

### 4.4.2 Equipment

The equipment for the melt processing of PVDF is the same as that for PVC or polyolefins, as during normal processing of PVDF, no corrosive products are formed. Thus no corrosion-resistant parts of the equipment are necessary.

### 4.4.3 Process Conditions

Extrusion temperatures vary between 230°C and 290°C (446°F and 554°F), depending on the equipment and the profile being extruded [4]. Sheet and cast film from slit dies are cooled on polished steel rolls kept at temperatures between 65°C and 140°C (149°F and 284°F). PVDF films can be monoaxially and biaxially oriented [4].

PVDF can be coextruded and laminated, but the process has its technical challenges in matching the coefficients of thermal expansion, melt viscosities, and layer adhesion. Special tie layers, often from blends of polymers compatible with PVDF, are used to achieve bonding [5,6]. As with any fluoropolymer, it is necessary to prevent long residence times in equipment and to purge the equipment after the process is finished. In addition, appropriate ventilation is necessary.

## 4.5 Production of PCTFE Films

### 4.5.1 Materials

The most common form of PCTFE are pellets that can be used in standard melt-processing techniques, including extrusion, injection molding, blow molding, and compression molding.

### 4.5.2 Equipment

Films are produced by extrusion using conventional extrusion equipment.

### 4.5.3 Process Conditions

PCTFE can be processed by most of the techniques used for thermoplastics. Processing temperatures for extrusion of PCTFE are typically in the range of 230°C–290°C (446°F–554°F). Since relatively high-molecular-weight resins are required for adequate mechanical properties, the melt viscosities are somewhat higher than those usual in the processing of thermoplastics. The reason is a borderline thermal stability of the melt, which does not tolerate sufficiently high processing temperatures [7]. Significant thermal decomposition of PCTFE occurs at temperatures above 300°C (570°F). Because of that, stringent temperature control is required to prevent degradation during extrusion.

## 4.6 Production of ECTFE Films

### 4.6.1 Materials

The most common form of ECTFE is hot-cut pellets that can be used in all melt-processing techniques, such as extrusion, injection molding, blow molding, compression molding, and fiber spinning [8].

### 4.6.2 Equipment

ECTFE is corrosive in its melt form; the surfaces of machinery that come in contact with the polymer must be lined with a highly corrosion-resistant alloy, for example, Hastelloy C-276. Certain grades with improved thermal stability and acid scavenging can be processed on conventional equipment [9].

### 4.6.3 Process Conditions

Also, as with any fluoropolymer, it is necessary to prevent long residence times in equipment and to purge the equipment after the process is finished. In addition, appropriate ventilation is necessary.

## 4.7 Production of THV Films

### 4.7.1 Materials

THV fluoroplastic resins are supplied as granules, agglomerate (coarse powder), and aqueous dispersion. As of this writing there are six grades in granular form and one of each as agglomerate and aqueous dispersion. The commercial resins have melting temperatures ranging from 125°C (257°F) to 225°C (437°F).

### 4.7.2 Equipment

Since the processing temperatures for THV resins are generally below the decomposition temperature of the polymers, there is no need to protect equipment against corrosion. Yet, as with any fluoropolymer, it is necessary to prevent long residence times in equipment and to purge the equipment after the process is finished. Also, appropriate ventilation is necessary.

### 4.7.3 Process Conditions

Generally, processing temperatures for THV are comparable to those used for most thermoplastics. In extrusion, melt temperatures at the die are in the 230°C–250°C (446°F–482°F) range. These relatively low processing temperatures open new options for combinations of different melts such as coextrusion, cross-head extrusion with thermoplastics as well as with various elastomers [10].

## 4.8 Production of Films From Other Thermoplastic Fluoropolymers Films

### 4.8.1 Production of Films From Fluorinated Thermoplastic Elastomers

#### *4.8.1.1 Materials*

*DAI-EL T-530* is supplied as translucent pellets with approximate melt temperature of 230°C (446°F). The polymer begins to decompose at temperature above 380°C (716°F) [11].

#### *4.8.1.2 Equipment*

Conventional extrusion equipment without corrosion-resistant components can be used as long as the processing temperature remains well under 380°C (716°F) [11].

#### *4.8.1.3 Process Conditions*

Processing temperatures used for DAI-EL T-530 are in the range of 240°C–280°C (464°F–536°F) at shear rates from 1 to $10^2$ second$^{-1}$ [12]. As with any fluoropolymer, it is necessary to prevent long residence times in equipment and to purge the equipment after the process is finished. Also, appropriate ventilation is necessary. To improve the strength of the products and to reduce their compression set at 150°C (302°F), they are irradiated by electron beam.

*Note*: Fluorinated thermoplastic elastomers of the TPV type are generally supplied as pellets and are processed using conventional equipment and processes. The process conditions depend on their composition, mainly on the type of the thermoplastic component used.

## 4.8.2 Production of Films From Amorphous Perfluoropolymers

### 4.8.2.1 Materials

*Teflon AF* is supplied as powder (two grades and as solution in special perfluorinated solvents). The powder forms are processed into films by conventional melt processing [13,14]. The solutions are processed into thin films by spin coating, spraying, and dip coating [15].

*CYTOP* is normally supplied in a 9 wt.% solution in perfluorinated solvents. The solutions are processed into thin films by spin coating, spraying, and dip coating [16]. Very thin films are producing by laser ablation and vacuum pyrolysis [17].

### 4.8.2.2 Equipment

Since Teflon AF begins to decompose above 360°C (680°F), the use of corrosion-resistant tooling is recommended for the melt extrusion of films [13]. As with any fluoropolymer, it is necessary to prevent long residence times in equipment and to purge the equipment after the process is finished. Also, appropriate ventilation is necessary.

### 4.8.2.3 Process Conditions

Typical processing temperature for Teflon AF 1600 is in the range of 240°C–275°C (464°F–527°F) and for Teflon AF 2400 in the range of 340°C–360°C (644°F–680°F). The powder grades can also be dissolved in perfluorinated solvents, for example, EC70 (3M Co.) [14]. As with any fluoropolymer, it is necessary to prevent long residence times in equipment and to purge the equipment after the process is finished. Also, appropriate ventilation is necessary.

## References

[1] S.V. Gangal, in: H.F. Mark, J. Kroschwitz (Eds.), Encyclopedia of Polymer Science and Technology, vol. 16, John Wiley & Sons, New York, 1989, p. 597.
[2] K. Hintzer, G. Löhr, in: J. Scheirs (Ed.), Modern Fluoropolymers, John Wiley & Sons, New York, 1997, p. 234.
[3] Extrusion Guide for Melt Processible Fluoropolymers, Bulletin E-85783, E. I. Du Pont de Nemours & Co., Wilmington, DE.
[4] J.E. Dohany, J.S. Humphrey, in: H.F. Mark, J. Kroschwitz (Eds.), Encyclopedia of Polymer Science and Technology, vol. 16, John Wiley & Sons, New York, 1989, p. 540.
[5] A. Stassel, US Patent 4,317,860, Produits Chimiques Ugine Kuhlmann, 1982.
[6] Y. Kitigawa, A. Nishioka, Y. Higuchi, T. Tsutsumi, T. Yamaguchi, T. Kato, US Patent 4,563,393, Japan Synthetic Rubber Co, Ltd, 1986.
[7] C.A. Sperati, in: I.I. Rubin (Ed.), Handbook of Plastic Materials and Technology, 1990, John Wiley & Sons, New York, 1990.
[8] G. Stanitis, in: J. Scheirs (Ed.), Modern Fluoropolymers, John Wiley & Sons, Ltd, Chichester, 1997, p. 528.
[9] G. Stanitis, in: J. Scheirs (Ed.), Modern Fluoropolymers, John Wiley & Sons, Ltd, Chichester, 1997, p. 529.
[10] D.E. Hull, B.V. Johnson, I.P. Rodricks, J.B. Staley, in: J. Scheirs (Ed.), Modern Fluoropolymers, John Wiley & Sons, Ltd, Chichester, 1997, p. 262.
[11] DAI-EL® T-350, Technical Data Sheet TDS-T-001 REV 0 05/03/16, Daikin America, 2016. <www.daikin-america.com>.
[12] M. Tatemoto, T. Shimizu, in: J. Scheirs (Ed.), Modern Fluoropolymers, John Wiley & Sons, Chichester, 1997, p. 571.
[13] P.R. Resnick, W.H. Buck, in: J. Scheirs (Ed.), Modern Fluoropolymers, John Wiley & Sons, Chichester, 1997, p. 417.
[14] J.G. Drobny, Technology of Fluoropolymers, second ed., CRC Press, Boca Raton, FL, 2009, p. 151.
[15] P.R. Resnick, W.H. Buck, in: J. Scheirs (Ed.), Modern Fluoropolymers, John Wiley & Sons, Chichester, 1997, p. 415.
[16] N. Sugiyama, in: J. Scheirs (Ed.), Modern Fluoropolymers, John Wiley & Sons, Chichester, 1997, p. 549.
[17] T.C. Nacson, J.A. Moore, T.M. Lu, Appl. Phys. Lett. 60 (1866).

# 5 Films from Polyvinyl Fluoride

The only known commercial supplier of polyvinyl fluoride (PVF) for the last 60 years has been the DuPont Company under the trademark Tedlar. DuPont produces a variety of films and powdered resins based on PVF [1]. The details on the production of the PVF resins are in Ref. [2]. PVF films are offered in both unoriented and oriented forms [3]. Unoriented PVF films are produced by casting onto a carrier web during which negligible stretching occurs [3]. The lack of orientation makes the film more formable and compliant than the oriented film. Oriented PVF films are produced by a method based on extrusion and orientation [3]. The orientation takes place by biaxial stretching that enhances the mechanical properties of the film in both machine and cross directions. Unoriented films, in addition to being formable, have lower tensile strength and higher elongation at break than the oriented ones. A comparison of the basic properties of 25 μm thick oriented and unoriented clear films is shown in Table 5.1 [4].

## 5.1 Properties of Polyvinyl Fluoride

PVF is a semicrystalline polymer with a planar, zigzag conformation [5,6]. The degree of crystallinity can vary significantly from 20% to 60% and is a function of defect structures. Commercial PVF is atactic, contains approximately 12% head-to-head linkages, and displays a peak melting point of about 190°C [7,8].

PVF displays several transitions below the melting temperature. Lower $T_g$ occurs at −15°C to −20°C and upper $T_g$ is in the range of 40°C–50°C. Two other transitions, at −80°C and 150°C, have been reported. PVF has low solubility in all solvents below about 100°C.

High-molecular-weight PVF is reported to degrade in an inert atmosphere, with concurrent hydrogen fluoride (HF) loss and backbone cleavage both occurring at about 450°C. In air, HF loss occurs at about 350°C, followed by backbone cleavage around 450°C.

PVF is transparent to radiation in the ultraviolet, visible, and near-infrared light regions, transmitting 90% of the radiation from 350 to

**Applications of Fluoropolymer Films.** DOI: https://doi.org/10.1016/B978-0-12-816128-9.00005-2

**Table 5.1 Comparison of Unoriented and Oriented Transparent Films at Room Temperature [4].**

| Property | Unoriented Film | Oriented Film |
|---|---|---|
| Ultimate tensile strength, MPa | 41 | 90 |
| Tensile module, MPa | – | 2075 |
| Elongation at break, % | 200 | 95 |
| Tear strength, initial, g/25 μm | 212 | 423 |

2500 nm. PVF becomes embrittled upon exposure to electron-beam radiation of 1000 Mrad but resists breakdown at lower doses. While PTFE is degraded at 0.2 Mrad (2 kGy), PVF retains its strength at 32 Mrad (320 kGy).

The self-ignition temperature of PVF film is 390°C. The limiting oxygen index for PVF is 22.6%. HF and a mixture of aromatic and aliphatic hydrocarbons are generated from the thermal degradation of PVF [9].

## 5.2 Processing Polyvinyl Fluoride

PVF is considered a thermoplastic, but it cannot be processed by conventional thermoplastic techniques because it is unstable above its melting point, which depending on composition ranges from 185°C to 195°C (365°F to 383°F) and already at 180°C (356°F) it produces a significant amount of hydrofluoric acid when heated in air [1]. The polymer contains a large number of hydrogen bonds and exhibits a high degree of crystallinity; therefore it is insoluble at room temperature in common organic solvents. However, some highly polar latent solvents, such as propylene carbonate, dimethylformamide, dimethylacetamide, butyrolactone, and dimethyl sulfoxide, dissolve it above 100°C (212°F) [1]. They also depress the melting point of PVF below the onset of accelerated degradation and below the boiling point of the solvent. The use of latent solvents is the basis of processes to manufacture PVF films and coatings. A latent solvent suitable for that has to have the appropriate volatility to allow the polymer particles to coalesce before complete evaporation.

## 5.3 Manufacturing of Unoriented Polyvinyl Fluoride Films

In this method the film is produced from a PVF dispersion in a latent solvent by a web-coating process using roll-coating technique [10]. The dispersion is applied to a moving carrier belt, the wet film levels on the carrying belt where it proceeds to the drying zone. The solvent is completely removed while the film coalesces. The solvent must have a sufficiently high boiling point to remain in the coating at least until the polymer begins to melt and film forming takes place. In the final step the dry film is removed (peeled) from the belt, which cycles back to the coating station [3].

The first step of the manufacturing process is the preparation of the dispersion. The PVF and optionally one or more dispersants and/or pigments are generally milled together. A wide variety of mills can be used for that operation [11]. The dispersion is milled for a time sufficient to de-agglomerate the PVF particles. Typical residence in an efficient production mill ranges from 30 seconds to 10 minutes [11]. The concentration of the dispersion will vary with the particular polymer, the process equipment, and the process condition [11]. In general, the amount of PVF in the dispersion is 30%–45% by weight.

The manufacturing process is depicted by Fig. 5.1 [12]. The figure shows the addition of PVF, solvent, additives, and pigments to the mix tank. In reality, PVF and pigment dispersions are prepared in advance and used to prepare the coating mixture. There are well-known

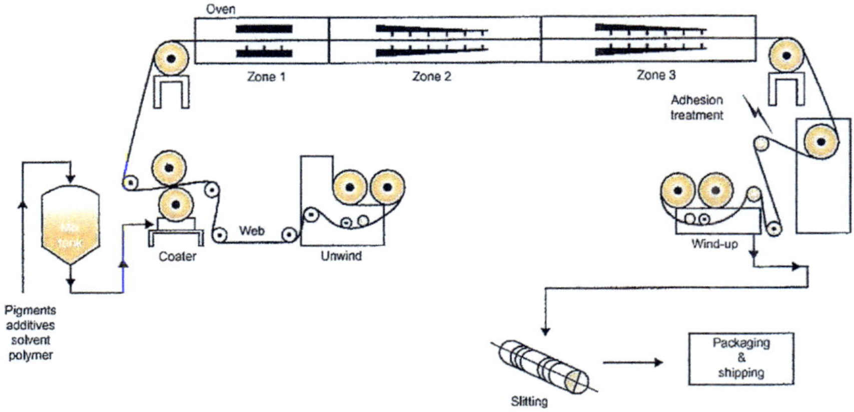


**Figure 5.1** Schematic diagram of the commercial manufacturing process of unoriented PVF film [12]. *PVF*, Polyvinyl fluoride.

methods for the preparation of pigments and polymers, detailed in Ref. [12]. The entire coating mix is pumped into a recirculated trough of the coater. A 25 μm thick polyester carrier web is fed into the coater from an unwind station. As mentioned earlier, the technology used to coat the web is a roll-coating technique, although its exact nature has not been disclosed. After passing through the coater the web enters a fairly long oven. The PVF mixture with latent solvent coalesces, which results in the formation of a film. During the remaining time, in the oven, the latent solvent is evaporated and the web is cooled prior to exit from the oven. Then it is either treated for adhesion online or off-line, followed by slitting and stripping the polyester web and packaging (Type 9 films) [13]. If the polyester carrier remains attached to the film, the product is designated as Type 8 film [14].

## 5.4 Manufacturing of Oriented Polyvinyl Fluoride Films

There are several steps in the preparation of oriented PVF films. The first step is the preparation of a homogeneous dispersion of PVF and pigments in a latent solvent. Next, the mixture is coalesced by heat in an extruder into a clear, thick, paste-like substance [15]. This step is followed by forcing the clear PVF paste through a slot die or a casting hopper into a thick film (sheet) and then promptly entering a quench bath followed by orientation and drying steps [15]. The entire process is discussed in detail in Ref. [15]. It should be noted that many details regarding this particular process are proprietary and/or not being published.

### 5.4.1 Polyvinyl Fluoride Dispersion in the Latent Solvent

PVF resin and solvent are mixed in a cylindrical vessel. This mixture usually contains agglomerates of polymer particles that have to be broken down into smaller size to allow the development of a good dispersion. This is accomplished by intense mixing of polymer particles and the solvent. In the large volume commercial operation, high-speed or high-shear dispersers or homogenizers are used. When pigments and other solid additives are required, the best choice is to add dispersions of the pigments and additives in the solvent prepared separately [15].

The mixture of PVF resin and solvent is passed through a homogenizer before being fed to the extruder [15]. Homogenizers work by forcing the feed through a small gap, thus dissipating large amounts of energy in the mixture. The energy breaks down the polymer agglomerates into smaller particles, thus improving the quality of the mixture.

### 5.4.2 Film Extrusion

As mentioned earlier, films from fluoroplastics are in general extruded by extrusion-casting or extrusion-blowing process. PVF films are manufactured by extrusion casting using a single-screw or twin-screw extruders [16]. The material is fed into the feed throat and is heated and transported through the machine and is shaped into a sheet by a *slot die* (see Fig. 5.2) [17] and then is moved immediately into a water bath where it is quenched [18]. The extrusion process results in film with a molecular orientation predominantly in the direction it is extruded, which is called *machine direction.* Properties of the film can be further significantly improved by orienting the film in the machine (longitudinal) and perpendicular directions also called *cross-machine* or *transverse* direction. A high purity of the melt, free of inclusions, is essential for film production. This is achieved by filtering the melt through a metal screen pack upstream of the film die.

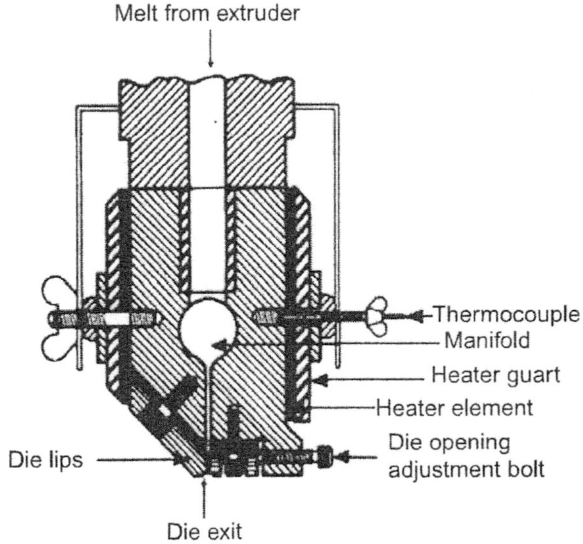


**Figure 5.2** Schematic of a typical slot die for cast film [17].

### 5.4.3 Biaxial Orientation

As mentioned earlier in Section 1.5.3, biaxial orientation (BO) is stretching a film or a sheet in both machine and transverse directions, which is carried out in a two-step process [18]. Here, in the case of PVF the extruded sheet, still containing the solvent, is stretched in the machine direction by passing it around heated rolls rotating at controlled and increasing speeds in excess of the extrusion speed. The degree of stretch is controlled by varying the roll speeds. After machine direction stretching is complete, the film enters the tenter frame for the transverse orientation.

As mentioned previously, in Chapter 1, Introduction, the film is during this process gripped by a transfixed clip system and moved forward while being stretched over the width. Tentering system is placed in a temperature-regulated tunnel. As the film passes through the tunnel, it is progressively stretched in the transverse direction as the clips diverge. Transverse stretch is controlled by varying the divergence of the edge-clip paths. During the passage, through the heated tunnel of the tenter frame, the solvent is removed from the oriented film [19]. After that, the film is wound up on a roll. The flow diagram of BO of PVF films is in Fig. 5.3 [20]. Later, the oriented film is fed into an adhesion treatment unit followed by slitting operation into the desired width. PVF films that do not require adhesion treatment (i.e., for release

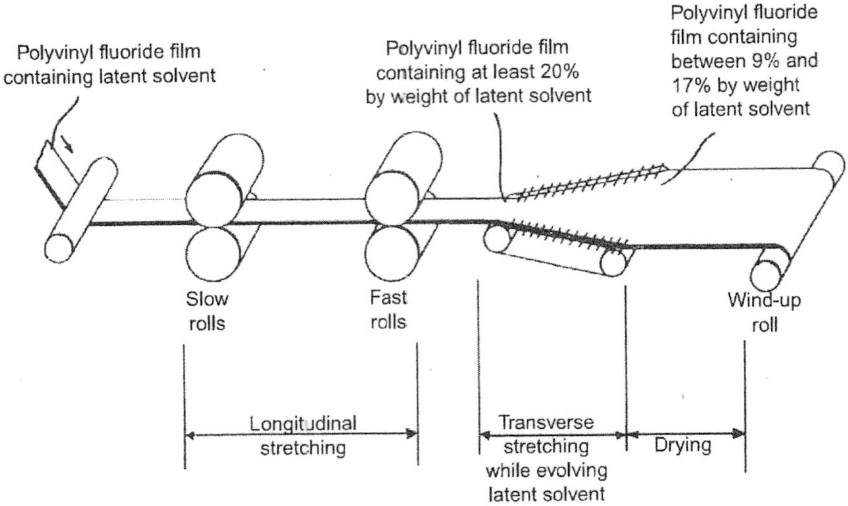


**Figure 5.3** Flow diagram of biaxial orientation of polyvinyl fluoride film [20].

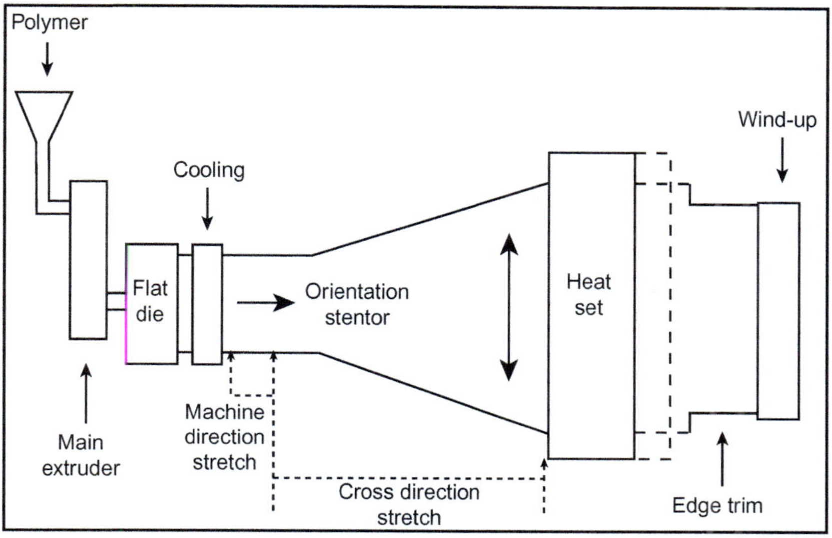


**Figure 5.4** Schematic diagram of tenter frame process biaxially oriented film [19].

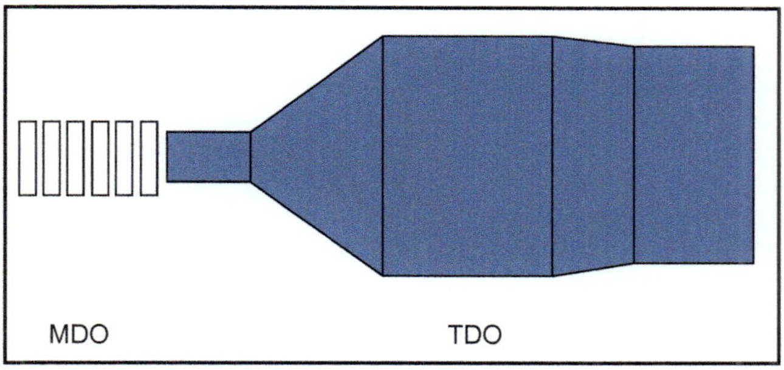


**Figure 5.5** Typical width profile of sequential orientation (MDO; TDO) [19]. *MDO*, Machine direction orientation; *TDO*, transverse direction orientation.

application) are sent to the slitting operation directly [17]. The overall schematic for the production of biaxially oriented PVF film is in Fig. 5.4 [20], and the typical width profile in this *sequential* orientation process is in Fig. 5.5 [20].

Mechanical properties of biaxially oriented films depend mainly on the amount of stretching and the ratio of the amount of individual orientations, an example of which is illustrated by Table 5.2 [21].

Schematic of a manufacturing process of commercial-oriented PVF film is depicted by Fig. 5.6 [22].

**Table 5.2 Mechanical Properties of Biaxially Oriented Polyvinyl Fluoride Films [21].**

| | Tensile Strength at Break, MPa | | Elongation at Break, % | | Tensile Modulus, MPa | | Tear Strength, g/mm | |
|---|---|---|---|---|---|---|---|---|
| Example No. | MD | TD | MD | TD | MD | TD | MD | TD |
| 1 | 151.7 | 96.6 | 75 | 132 | 2207 | 1862 | – | – |
| 2 | 100 | 93 | 105 | 115 | 1862 | 1793 | 800 | 880 |
| 3 | 124.1 | 124.1 | 99 | 96 | 2000 | 2069 | 400 | 360 |

*MD*, Machine direction; *TD*, transverse direction.

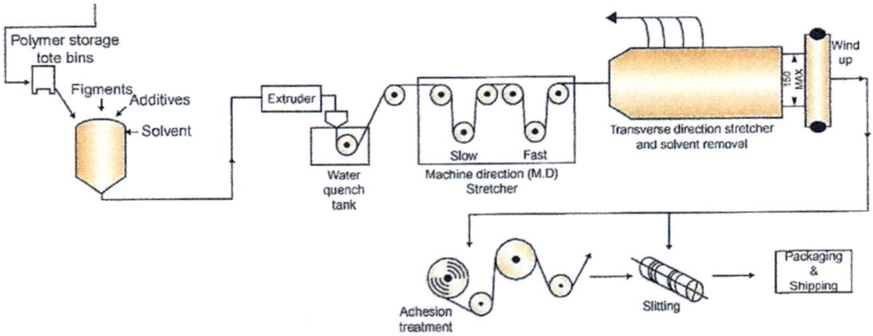


**Figure 5.6** Schematic diagram of the commercial manufacturing process of oriented PVF film [22]. *PVF*, Polyvinyl fluoride.

## References


[1] S. Ebnesajjad, Fluoroplastics, second ed., Melt Processible Fluoropolymers, vol. 2, Elsevier, Oxford, UK, 2015, p. 319.

[2] S. Ebnesajjad, Polyvinyl Fluoride, Technology and Applications of PVF, Elsevier, Oxford, UK, 2013.

[3] S. Ebnesajjad, Polyvinyl Fluoride, Technology and Applications of PVF, Elsevier, Oxford, UK, 2013, p. 141.

[4] S. Ebnesajjad, Polyvinyl Fluoride, Technology and Applications of PVF, Elsevier, Oxford, UK, 2013, p. 142.

[5] D.P. Carlson, US Patent 3,536,733 (Oct. 27, 1970) to DuPont Co.

[6] H.S. Eleuterio, G.W. Meschke, US Patent 3,358,003 (Dec. 12, 1967) to DuPont Co.

[7] Englander F, Meyer G, US Patent 3,987,117 (Oct. 19, 1976) to Dynamit Nobel.

[8] S. Ebnesajjad, Kirk-Othmer Encyclopedia of Chemical Technology, fourth ed., John Wiley & Sons, New York, 2004.

[9] S. Ebnesajjad, Introduction to fluoropolymers, in: M. Kutz (Ed.), Applied Plastics Engineering Handbook, second ed., Elsevier, Oxford, UK, 2016.

[10] S. Ebnesajjad, Polyvinyl Fluoride, Technology and Applications of PVF, Elsevier, Oxford, UK, 2013, p. 144.

[11] S. Ebnesajjad, Polyvinyl Fluoride, Technology and Applications of PVF, Elsevier, Oxford, UK, 2013, p. 146.

[12] S. Ebnesajjad, Polyvinyl Fluoride, Technology and Applications of PVF, Elsevier, Oxford, UK, 2013, p. 148.

[13] S. Ebnesajjad, Polyvinyl Fluoride, Technology and Applications of PVF, Elsevier, Oxford, UK, 2013, p. 149.

[14] S. Ebnesajjad, Polyvinyl Fluoride, Technology and Applications of PVF, Elsevier, Oxford, UK, 2013, p. 145.

[15] S. Ebnesajjad, Polyvinyl Fluoride, Technology and Applications of PVF, Elsevier, Oxford, UK, 2013, p. 119.

[16] S. Ebnesajjad, Polyvinyl Fluoride, Technology and Applications of PVF, Elsevier, Oxford, UK, 2013.

[17] S. Ebnesajjad, Polyvinyl Fluoride, Technology and Applications of PVF, Elsevier, Oxford, UK, 2013, p. 122.

[18] S. Ebnesajjad, Polyvinyl Fluoride, Technology and Applications of PVF, Elsevier, Oxford, UK, 2013, p. 123.

[19] S. Ebnesajjad, Polyvinyl Fluoride, Technology and Applications of PVF, Elsevier, Oxford, UK, 2013, p. 124.

[20] S. Ebnesajjad, Polyvinyl Fluoride, Technology and Applications of PVF, Elsevier, Oxford, UK, 2013, p. 130.

[21] S. Ebnesajjad, Polyvinyl Fluoride, Technology and Applications of PVF, Elsevier, Oxford, UK, 2013, p. 135.

[22] S. Ebnesajjad, Polyvinyl Fluoride, Technology and Applications of PVF, Elsevier, Oxford, UK, 2013, p. 138.

# 6 Secondary Processing of Fluoropolymer Films

Thermoplastics films are often subjected to additional operations to expand their utility as discussed in Section 1.3. Since the fluoropolymer films discussed here are produced from thermoplastics (fluoroplastics), many secondary processing and fabrication methods being used for them are identical or similar. However, some of them differ. The subject is discussed in this chapter.

## 6.1 Surface Preparation of Fluoropolymer Films

Fluoropolymers have inherently lower surface energy than most other polymers and because of that they tend to form poor adhesive bonds without any type of surface treatment. There are many physical and chemical treatments used for that purpose described in Section 1.3.1. Industrial surface treatments for plastics include corona, flame and plasma treatment, and chemical etching. The increase in surface energy of fluoropolymers, surface modification, involves significant dehalogenation, that is, the removal of fluorine and chlorine atoms from the surface of the macromolecules.

### 6.1.1 Corona Treatment of Fluoropolymer Films

Corona is a stream of charged particles that are accelerated by an electric field. It is generated when a space gap filled with air or other gases is subjected to sufficiently high voltage in order to set up a chain reaction of high-velocity particle collision with neutral molecules, resulting in the generation of more ions. It is reported that corona treatment is used for FEP and PTFE films and polyvinyl fluoride (PVF) (see Section 1.3.1.1) although other fluoroplastic films could be treated by corona when necessary.

### 6.1.2 Sodium Etching of Fluoropolymer Films

Perfluorinated fluoroplastics are chemically unaffected by nearly all commercial chemicals with the exception of highly oxidizing substances as elemental forms of sodium, potassium, and other alkaline metals. This is the basis for sodium etching of fluoroplastic parts. The original

**Applications of Fluoropolymer Films.** DOI: https://doi.org/10.1016/B978-0-12-816128-9.00006-4

method for surface treatment for adhesive bonding was developed for PTFE and was etched by a sodium solution in anhydrous liquid ammonia. An alternative solution of sodium is to prepare a complex with naphthalene followed by dissolution in tetrahydrofuran or dimethyl glycol ether. Newer systems using glycol diethers (referred to as *glymes*) are much less toxic than tetrahydrofuran [1].

Special precautions must be taken while working with sodium etching solutions. Fluoropolymer products (films, parts), treated by sodium etching solution, should be stored in a cold, dark atmosphere free from oxygen and moisture. The useful shelf life of etched polymer under these conditions at temperature lower than 5°C is 3–4 months.

For in-house etching, it is advisable to consider purchasing commercial etching solutions available from a number of sources. Some of them can be used successfully for fluoropolymers other than PTFE. Moreover, there are several companies offering full service contract etching.

### 6.1.3 Plasma Treatment of Fluoropolymer Films

Plasma is considered the fourth state of the matter and is produced by exciting a gas with electrical energy introduced into a vacuum chamber. Since plasma is intensely reactive, it can effectively modify surfaces of plastics. It can be used to treat thermoplastic films to impart hardness, roughness, more or less wettability, and increased adherability to the part surfaces. Plasma treatment oxidizes the surface of the polymer in the presence of oxygen, which is believed to be the reason for roughening of the surface.

*Atmospheric plasma treatment* (APT), also called *glow discharge*, is operating without the use of vacuum. The plasma creates uniform *plasma cloud* that completely surrounds small objects or spreads into the boundary layer of the surface. Fluoropolymer films can be treated by APT with PVF responding to it well but PTFE does not as pointed out in Section 1.3.1.2.

### 6.1.4 Flame Treatment of Fluoropolymer Films

Flame treatment is defined as a surface-preparation technique in which the plastic is briefly exposed to a flame. Flame treatment oxidizes the surface through a free radical mechanism, introducing hydroxyl, carbonyl, and amide functional group to the depth of about 4–6 nm, and produces chain scissions and some cross linking. The film is passed over an oxidizing flame formed by an oxygen-rich (relative to stoichiometry) mixture of hydrocarbon gases (see Section 1.3.1.3). However, this method is not effective in the adhesion treatment of perfluoroplastics, but it is reported that it can be used for PVF and ECTFE [2].

## 6.2 Lamination of Fluoropolymer Films

Lamination is the technique of manufacturing a material in multiple layers, so that the composite material achieves improved strength, stability, sound insulation, appearance, or other properties from the use of differing materials. Details are given in Section 1.3.2. In extrusion lamination the molten polymer bonds to a solid substrate by melting chemical interaction and in some cases by mechanical interlocking [3]. Fluoropolymer films can be laminated by methods mentioned in Section 1.3.2 and by extrusion lamination. The most widely used lamination is used in the processing of PVF films. PVF films are laminated to metals including aluminum, stainless steel, cold-rolled steel, galvanized steel, copper, and titanium as well as to other substrates, such as cellulosic materials; to films and sheets from other polymers to fiberglass panels; and to vinyl wall coverings. In each case the other substrate has to be cleaned and/or surface treated, and in most of the cases, special adhesives are used. The best current source of information on this subject is Ref. [4].

## 6.3 Heat Sealing of Fluoropolymer Films

Heat sealing is a widely used process in the packaging industry. It involves joining two polymer films by the application of heat and pressure for a specified time. As pointed out in Section 1.3.3, the polymer characteristics that play a critical role in the heat-sealing process involve melting temperature, chain diffusion rate, melt strength, and crystallization kinetics. Thermoplastic films, including fluoroplastic films, can be sealed by any method that heats the contacting surfaces of the film above the melting point of the given polymer and at the same time. Applied pressure assures intimate contact of those surfaces. Even here, the techniques usually used are hot bar heat sealing, impulse heat sealing, and hot air sealing.

## 6.4 Metallization of Fluoropolymer Films

Plastic parts can be coated with metal, in a process called metallization, for both esthetic and mechanical purposes. Metalized plastic components are used in similar applications as metal plated parts, but tend to be lower in weight and have higher corrosion resistance, although not in all cases. As pointed out in Section 1.3.4, fluoropolymer films (PTFE, FEP, ETFE, ECTFE, PVDF, and PVF) can be readily metallized by a variety of metals by vacuum and electroless deposition. Aluminum-coated PTFE is used in high temperature–resistant capacitors. Good adhesion of fluoropolymers to copper has been achieved by silanization of fluoropolymer film surfaces. Preactivation of PTFE surface improves adhesion of copper applied by electroless deposition of copper [5]. More information on metallization of fluoropolymers is in Ref. [5].

## 6.5 Orientation of Fluoropolymer Films

As pointed out in Section 1.3.5, polymeric films usually emerge from melt simple-processing methods such as extrusion and calendering with the polymer chains arranged in a relatively random order, because of which they exhibit *anisotropy*. If the film is stretched, the polymer chains tend to line up (or orient) in the direction of the stretch. Consequently, this orientation affects the physical properties of the film in the direction of the stretch, referred to as machine direction orientation and transverse direction orientation that is stretching the film across the process flow. Another possibility is the stretching in both directions, which is referred to as biaxial orientation.

Essentially any thermoplastic film can be oriented by these methods. As for fluoropolymer films, mainly PVF, PVDF and PTFE are biaxially oriented on industrial scale. PCTFE alone cannot be oriented because of its high degree of crystallinity but when combined into laminated with other polymeric films, it is possible [6,7]. More details on orientation of fluoropolymer films are in Ref. [5].

## 6.6 Other Methods for Secondary Processing of Fluoropolymer Films

### 6.6.1 Laser Marking of Fluoropolymer Films

Laser marking is the modification of the surface of a material to create human- and/or machine-readable alphanumeric characters, 1D/2D barcodes, logos, insignia, etc., mainly for the purpose of identification.

Laser marking of fluoropolymer films can be performed with either a 10.6 or 9.3 μm $CO_2$ laser, with no appreciable difference in process quality. Laser marking of pigmented fluoropolymer substrates is enhanced by using titanium dioxide pigment coated with organosilane [8]. In general, laser marking can be performed on PTFE, perfluoroalkoxy, FEP, and ETFE [9].

### 6.6.2 Laser Cutting of Fluoropolymer Films

Laser cutting is the complete removal and separation of material from the top to the bottom surface along a designated path. Laser cutting can be performed on a single layer material or multilayer material. Material thickness and density are important factors to consider when laser cutting. Cutting through thin material requires less laser energy than cutting the same material in a thicker form. Lower density material typically requires less laser energy. However, increasing laser power level generally improves laser cutting speed. Laser cutting of fluoropolymer films can also be performed with either a 10.6- or 9.3-μm $CO_2$ laser [8].

### 6.6.3 Printing on Fluoropolymer Films

Printing on fluoropolymer films requires surface treatment as described in Section 6.1 and often the use of a special primer [10]. The high-speed roll-to-roll techniques used for printing on fluoropolymer films are *gravure printing* and *flexographic printing* [11].

The limitation for inks used to mark fluoropolymer substrates is that they have to be able to withstand high curing temperatures, as this is the adhesion mechanism. This limits the selection of the colorants to very heat-stable pigments [12].

But because the melting point of ECTFE, ETFE, and PVDF are lower than those of other commonly used fluoropolymers, a lower curing temperature material can be used. When the material is cured properly, the print will melt into the top layer of the surface of an application. This provides a very good adhesion to the fluoropolymer, and an extremely long-lasting mark. The printing and striping inks used with PTFE and FEP are formulated from the same resins as the substrate. The use of the same matrix material creates a uniformly melted top layer [12].

### 6.6.4 Thermoforming of Fluoropolymer Films

Single films, sheets, and laminates from fluoropolymer films can be formed into required shapes by a process, referred to as *thermoforming*. Thermoforming is a manufacturing process in which a plastic film, sheet, or laminate is heated to a pliable forming temperature, formed to a specific shape in a mold, and trimmed to create a useable product. The film, sheet, or laminate is heated in an oven to a temperature high

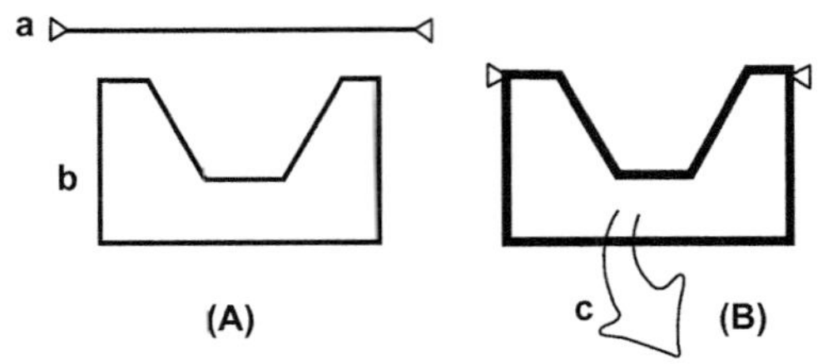


**Figure 6.1** Vacuum forming [13].
(A) Preheated sheet prior to forming
(B) Formed sheet into female mold
a—Preheated, clamped sheet
b—Female mold with vacuum holes
c—Vacuum

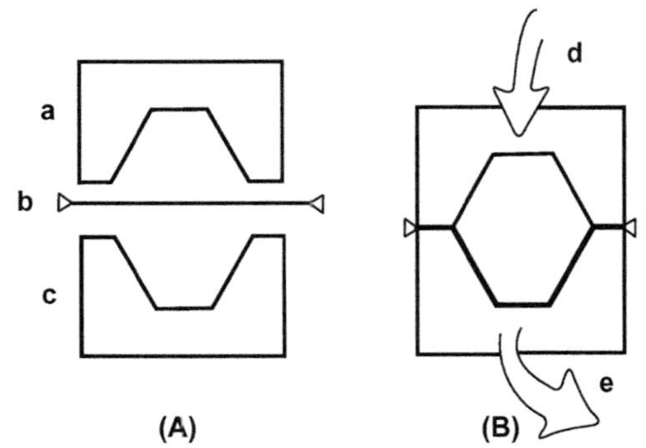


**Figure 6.2** Pressure forming [13].
(A) Preheated sheet prior to forming
(B) Formed sheet into female mold
a—Pressure box
b—Preheated, clamped sheet
c—Female mold with vacuum/vent holes
d—Applied air pressure
e—Venting of vacuum

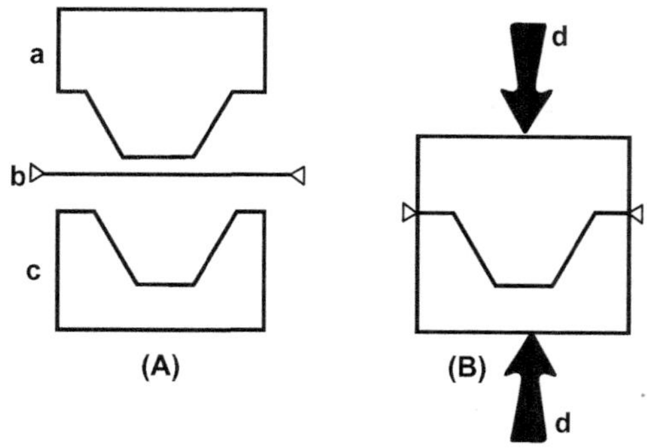


**Figure 6.3** Matched mold (die) forming [13]
(A) Preheated sheet for forming.
(B) Sheet formed by simultaneous motion of two mold halves
a—Male mold half
b—Preheated, clamped sheet
c—Female mold half
d—Applied force.

enough so that it can be stretched into or onto a mold and cooled to a finished shape [13]. Currently, there are three thermoforming methods, namely, *vacuum forming*, *pressure forming*, and *match die forming*. The principles of these methods are shown in Figs. 6.1–6.3.

## References

[1] S. Ebnesajjad, Chapter 8 Polyvinyl Fluoride—Technology and Applications of PVF, Elsevier, Oxford, UK, 2013.
[2] S. Ebnesajjad, Polyvinyl Fluoride—Technology and Applications of PVF, Elsevier, Oxford, UK, 2013, p. 204.
[3] B.A. Morris, The Science and Technology of Flexible Packaging, Elsevier, Oxford, UK, 2017, p. 371.
[4] S. Ebnesajjad, Chapter 9 Polyvinyl Fluoride—Technology and Applications of PVF, Elsevier, Oxford, UK, 2013.
[5] M. Friedman, G. Walsh, Polym. Eng. Sci. 42 (8) (2002) 1768.
[6] M.L. Tsai, US Patent 5,874,035 (February 23, 1999) to Allied Signal, Inc.
[7] M.L. Tsai, US Patent 5,945,221 (February 23, 1999) to Allied Signal, Inc.
[8] Laser Processing, Universal Laser Systems, <www.ulsinc.com/materials/fluoropolymers>, 2019.
[9] Laser Marking, Zeus Industrial Products, <www.zeusinc.com/products/value-add/laser-marking>, 2019.

[10] S. Ebnesajjad, Polyvinyl Fluoride—Technology and Applications of PVF, Elsevier, Oxford, UK, 2013, p. 214.
[11] B.A. Morris, The Science and Technology of Flexible Packaging, Elsevier, Oxford, UK, 2017, p. 46.
[12] Considerations When Printing and Marking Fluoropolymers, PolyOne Corporation, <www.polyone.com>, 2015.
[13] S. Ebnesajjad, Polyvinyl Fluoride—Technology and Applications of PVF, Elsevier, Oxford, UK, 2013, p. 251.

# 7 Testing of Thermoplastic Films

There is a multitude of material standards that govern the basic properties and performance of thermoplastic films and sheeting. These standards account for everything from simple, single-material films to multimaterial, anisotropic constructions and help to characterize such properties as tensile strength, tearing resistance, impact resistance, folding resistance, puncture energy, and friction coefficients. Additional standards apply to the testing of thermal, electrical, barrier and optical properties, chemical resistance, aging resistance, as well as other properties and performance pertaining to their practical applications. Essentially, all the test procedures discussed here apply to fluoroplastics that are used for the manufacture of films and sheets as well as of the final products.

## 7.1 Standards for Testing of Mechanical Properties

**ASTM D618**: Standard Practice for Conditioning Plastics for Testing
**ASTM D638**: Standard Test Method for Tensile Properties of Plastics
**ASTM D790**: Standard Tests Methods for Flexural Properties of Unreinforced and Reinforced Plastics and Insulating Materials
**ISO 178**: Plastics—Determination of flexural properties
**ASTM D792**: Standard Test Methods for Density and Specific Gravity (Relative Density) of Plastics by Displacement
**ASTM D882**: Tensile Properties of Thin Plastic Sheeting
**ASTM D1004**: Standard Method for Tear Resistance (Graves Tear) of Plastic Film and Sheeting
**ASTM D1708**: Tensile Properties of Plastics by Use of Microtensile Specimens
**ASTM D1709**: Impact Resistance of Plastic Film by the Free-Falling Dart Method
**ASTM D1894**: Static and Kinetic Coefficients of Friction of Plastic Film and Sheeting
**ASTM D1922**: Propagation Tear Resistance of Plastic Film and Thin Sheeting by Pendulum Method

**Applications of Fluoropolymer Films.** DOI: https://doi.org/10.1016/B978-0-12-816128-9.00007-6

**ASTM D1938**: Standard Test Method for Tear-Propagation Resistance (Trouser Tear) of Plastic Film and Thin Sheeting by a Single-Tear Method.
**ISO 6583-2**: Plastics-Film and Sheeting-Determination of Tear Resistance—Part2: Elmendorf Method
**ASTM D2176**: Standard Test Method for Folding Endurance of Paper and Plastic Film by the M.I.T. Tester
**ASTM D3420**: Pendulum Impact Resistance of Plastic Film
**ASTM D7192**: Standard Test Method for High Speed Puncture Properties of Plastic Films Using Load and Displacement Sensors.

## 7.2 Standards for Testing of Thermal Properties

### 7.2.1 Heat Aging

**ASTM D3045**: Standard Practice for Heat Aging of Plastics without Load
**ISO 2578**: Determination of Time-Temperature Limits After Exposure to Prolonged Action of Heat

### 7.2.2 Thermal Stability

**ASTM E2550**: Standard Test Method for Thermal Stability by Thermogravimetry

### 7.2.3 Heat Sealability

**ASTM F2029**: Standard Practices for Making Laboratory Heat Seals for Determination of Heat Sealability of Flexible Barrier Materials as Measured by Seal Strength

### 7.2.4 Shrinkage

**ASTM D2732**: Standard Test Method for Unrestrained Linear Thermal Shrinkage of Plastic Film and Sheeting

### 7.2.5 Linear Coefficient of Expansion

**ASTM D696**: Standard Test Method for Coefficient of Linear Thermal Expansion of Plastics between −30°C and 30°C with a Vitreous Silica Dilatometer

### 7.2.6 Continuous Use Temperature

#### *7.2.6.1 Long Cycle Use UL 746B*

Standard for Polymeric Materials—Long Term Property Evaluations.

#### *7.2.6.2 Short Cycle Use UL 746A*

Standard for Polymeric Materials—Short Term Property Evaluations.

## 7.3 Standards for Testing of Electrical Properties

**ASTM D149**: Standard Test Method for Dielectric Breakdown Voltage and Dielectric Strength of Solid Electrical Insulating Materials at Commercial Power Frequencies
**ASTM D150**: Standard Test Methods for AC Loss Characteristics and Permittivity (Dielectric Constant) of Solid Electrical Insulation
**ASTM D257**: Standard Test Methods for DC Resistance or Conductance of Insulating Materials
**ASTM D495**: Standard Test Method for High-Voltage, Low-Current, Dry Arc Resistance of Solid Electrical Insulation

## 7.4 Standards for Testing of Barrier Properties

**ASTM F1306**: Slow Rate Penetration Resistance of Flexible Barrier Films and Laminates
**ASTM D3985**: Oxygen Permeation of Flexible Materials Using a Coulometric Sensor
**ASTM E96**: Water Vapor Transmission of Materials
**ASTM F1249**: Standard Test Method for Water Vapor Transmission Rate through Plastic Film and Sheeting Using Modulated Infrared Sensor
**ISO 15105-2**: Plastics-Film and sheeting—Determination of gas transmission rate—Part 2: Equal pressure method
**ISO 1506-2**: Plastics-Film and Sheeting-Determination of Water Vapour Transmission Rate—Part 2: Infrared detection sensor method

## 7.5 Standards for Testing of Optical Properties

### 7.5.1 Refractive Index

**ASTM D542**: Standard Test Method for Index of Refraction of Transparent Organic Plastics

### 7.5.2 Optical Transmission

**ASTM E424**: Standard Test Methods for Solar Energy Transmittance and Reflectance (Terrestrial) of Sheet Materials
**ASTM D1003**: Standard Test Method for Haze and Luminous Transmittance of Transparent Plastics

## 7.6 Standards for Testing of Chemical Resistance

**ASTM D543**: Standard Practices for Evaluating the Resistance of Plastics to Chemical Reagents

## 7.7 Standards for Testing of Surface Properties

### 7.7.1 Critical Surface Energy

**ISO-19403-1**: Paints and varnishes—Wettability—Part 1: Terminology and general principles

### 7.7.2 Contact Angle

**ASTM D7334**: Standard Practice for Surface Wettability of Coatings, Substrates and Pigments by Advancing Contact Angle Measurement

## 7.8 Standards for Testing of Thermal Stability

**ASTM E2250**: Standard Test Method for Thermal Stability by Thermogravimetry

## 7.9 Standards for Testing of Weatherability

### 7.9.1 Accelerated Testing Methods

**ASTM G154**: Standard Practice for Operating Fluorescent Ultraviolet (UV) Lamp Apparatus for Exposure of Nonmetallic Materials
**ASTM G155**: Standard Practice for Operating Xenon-Arc Light Apparatus for Exposure of Nonmetallic Material
**ASTM D4329**: Standard Practice for Fluorescent Ultraviolet (UV) Lamp Apparatus Exposure of Plastics
**ISO 4892-2**: Plastics—Methods of exposure to laboratory light sources—Part 2: Xenon-arc amps

### 7.9.2 Outdoor Testing Methods

**ASTM D1435**: Standard Practice for Outdoor Weathering of Plastics
**ISO 877-2**: Plastics—Methods of exposure to solar radiation—Part 2: Direct weathering and exposure behind glass

More detailed discussion of this subject is in Refs. [1,2].

## References

[1] L.W. McKeen, Chapter 2 Film Properties of Plastics and Elastomers, fourth ed., Elsevier, Oxford, 2017.
[2] B.A. Morris, The Science and Technology of Flexible Packaging, Part IV: Film Properties, Elsevier, Oxford, 2017.

# 8 Safety, Hygiene, Disposal, Recycling of Fluoropolymer Films

## 8.1 Safety, Hygiene, and Disposal of Fluoropolymer Films

### 8.1.1 Toxicology of Fluoroplastics

Fluoroplastics are chemically very stable and inert. However, they can produce toxic compounds if overheated. Precautions should be taken to remove any degradation fragments produced during the processing and fabrication of parts from them. Filled or compounded resins contain pigments, surfactants, and other additives to modify the properties of the polymers. These additives may present some hazards in the processing of the compounded resins. For example, aqueous dispersions of fluoropolymers frequently contain surfactants that may produce adverse physiological symptoms. Such hazards should be considered by themselves and in conjunction with fluoropolymers [1]. Safety information provided by manufacturers of the additives and the compounds should be consulted.

### 8.1.2 Thermal Behavior of Fluoroplastics

During usual processing, fluoroplastics are heated to high temperatures and degraded to some extent. The extent of degradation and the type of degradation products depend on variables, such as temperature, presence of oxygen, physical form of the article, duration of exposure to the temperature, and presence of additives. The products of decomposition of fluoropolymers fall into three categories: fluoroalkenes, oxidation products, and particulates of low molecular weight fluoropolymers. A major product of oxidation of PTFE is carbonyl fluoride that is highly toxic and hydrolyzes into hydrofluoric acid and carbon dioxide. At 450°C (842°F) in the presence of oxygen, PTFE degrades into carbonyl fluoride and hydrofluoric acid. At 800°C (1472°F), tetrafluoromethane is formed [1]. It is important to follow the recommendations and specifications of the

**Applications of Fluoropolymer Films.** DOI: https://doi.org/10.1016/B978-0-12-816128-9.00008-8

**Table 8.1 Maximum Continuous Use and Processing Temperatures of Common Fluoroplastics.**

| Polymer | Maximum Continuous Use Temperature (°C) | Typical Processing Temperature (°C) |
|---|---|---|
| PTFE | 260 | 380 |
| PFA | 260 | 380 |
| FEP | 205 | 360 |
| ETFE | 150 | 310 |
| ECTFE | 150 | 280 |
| PCTFE | 120 | 265 |
| PVF | 107 | 193 |
| PVDF | 120 | 230 |

suppliers of resins and parts. The Guide to Safe Handling of Fluoropolymer Resins, published by the Society of Plastics Industry, Inc. (see Table 8.1), specifies the maximum continuous use and processing temperatures for selected fluoroplastics [2].

Operation of processing equipment at high temperatures may result in the generation of toxic gases and particulate fume. The best known adverse effect on humans is *polymer fume fever.* This temporary flu-like condition, lasting typically 24 hours causes fever, chills, and occasionally coughs. It is further enhanced by tobacco smoking. It has been suggested that no health hazards exist, if the fluoroplastics are heated at temperatures below 300°C (572°F) [3]. Details on health hazards of decomposition products of fluoropolymers are in Refs. [2,3]. These risks prompted the establishment of exposure limits by various regulatory agencies (see Table 8.2) [4].

There are a number of measures that can be taken to reduce and control the exposure of personnel to monomers and decomposition products during the processing of fluoroplastics. These include ventilation, spillage cleanup, equipment cleanup, proper maintenance, and elimination of fire hazard. The personnel involved in the processing should wear protective clothing, maintain strict personal hygiene, and be made aware of incompatibility of specific materials [5].

Processing and fabrication to finished products may be hazardous, because often very high processing temperatures are used. Depending

**Table 8.2** **Exposure Limit Types.**

| Limit | Type | Source |
|---|---|---|
| PEL | Legal OEL | US Code of Federal Regulations, Title 29, Part 1910 (29 CFR 1910) |
| TLV | REL | American Conference of Governmental Industrial Hygienists or NIOSH |

*NIOSH*, National Institute for Occupational Safety and Health; *OEL*, occupational exposure limit; *PEL*, permissible exposure limit; *REL*, recommended exposure limit; *TLV*, threshold limit values.
*Source*: Data from The Society of Plastics Industry, The Guide for Safe Handling of Fluoropolymer Resins, fourth ed., The Society of Plastics Industry, Washington, DC, 2005.

on a specific process, fumes may not be present in the amounts to affect the personnel, but protection, such as protective clothing and gloves, should be used [5].

### 8.1.3 Medical Applications of Fluoroplastics

Because of their inertness and relative purity, some fluoroplastics, for example, certain grades of PTFE, are used in medical applications. In such cases, Federal Food and Drug Administration most often reviews and approves the entire medical device, not its components, such as specific parts or resin used. Resin suppliers have strict specific policies about the use of their resins in medical devices [5].

### 8.1.4 Food Contact

Fluoropolymer resins are covered in the United States by Federal Food, Drug and Cosmetic Act, 21 CFR & 177.1380 & 177.1550 and in the European Union by the EC Directive 90/128 [5]. Several fluoroplastics, such as PTFE, PFA, and FEP, have been approved by FDA for contact with food. Additives, such as stabilizers, antioxidants, pigments, and others, must be approved to meet the food-additive regulations if they have no prior clearance [5].

### 8.1.5 Environmental Protection and Disposal Methods for Fluoroplastics

The preferred methods for disposing fluoroplastics are recycling and reusing them. This subject is covered in more detail in Section 8.2. Landfilling of fluoroplastics is permitted in some cases by local regulations because they are environmentally stable and contain no harmful substances. This is justified if there is no valid recycling or incineration. When disposing suspensions and dispersions, solids should be removed from the liquids and then disposed off. Incineration can be used only when the incinerator is equipped by a scrubber for removing hydrogen fluoride, hydrogen chloride, or other acidic products of combustion. If the scrap contains pigments, additives, and solvents, it should be handled in a manner to meet local regulations for nonfluoropolymer ingredients. Some of the mixtures may require compliance with local regulations for hazardous materials [6].

## 8.2 Recycling of Fluoropolymer Films

As with any raw material, recycling of fluoropolymers is very important. Most melt-processible fluoropolymers can be reprocessed in a fashion similar to other thermoplastics. With PTFE the situation is more complicated. Because of its high melt viscosity, it is difficult to remelt and mix with virgin material, particularly if it contains mineral fillers. Nevertheless, a significant amount of PTFE production scrap is being reused by cleaning, grinding, and using in that form for ram extrusion [7].

Currently, most PTFE scrap (mainly residues from machining operations) is processed by radiation, being exposed to doses up to 400 kGy to reduce the molecular weight drastically and to obtain micropowders [8]. The most common process employs an electron-beam processor, although gamma radiation can also be used. High-molecular-weight PTFE can also be converted into micropowders by thermal or shear degradation [8]. A process involving chemical recycling of PTFE using fluidized bed has been developed and patented [9,10].

The optimum temperature is in the range of 545°C–600°C (1013°F–1112°F), and the main decomposition products are tetrafluoroethylene (TFE), hexafluoropropene (HFP), and cyclo-perfluorobutane ($c\text{-}C_4F_8$). The most important advantages of this process are that the monomers produced can be purified before repolymerization, which allows the production of a more valuable product, and that the process is continuous [9]. Another continuous process is the pyrolysis of PTFE scrap in a reactor heated by radiofrequency induction to 600°C–900°C (1112°F–1652°F). The products are TFE, HFP, and c-$C_4F_8$. The yield is in excess of 94% [11].

The use of reprocessed (reground) PTFE resin has limitations. For one thing, it exhibits markedly lower tensile strength and elongation than virgin PTFE. Reprocessed material creeps up to 25% more than virgin resin and contains twice the void content [12]. Because of porosity and larger number of voids, its dielectric strength is lower than that of the virgin PTFE. Thus the use of reprocessed material is limited to such applications, where cost is an important consideration and lower performance is sufficient for the application.

## References

[1] S. Ebnesajjad, P.R. Khaladkar, Fluoropolymers Applications in Chemical Processing Industries, William Andrew, Inc, Norwich, NY, 2005, p. 385.
[2] The Society of Plastic Industry, Inc., The Guide to Safe Handling of Fluoropolymers Resins, The Society of Plastic Industry, Inc., 1988.
[3] C.A. Rose, Environmental and Occupational Medicine, second ed., Little, Brown and Co., Boston, 1992, p. 373.
[4] The Society of Plastics Industry, The Guide for Safe Handling of Fluoropolymer Resins, fourth ed., The Society of Plastics Industry, Washington, DC, 2005.
[5] S. Ebnesajjad, P.R. Khaladkar, Fluoropolymers Applications in Chemical Processing Industries, William Andrew, Inc, Norwich, NY, 2005, p. 391.
[6] S. Ebnesajjad, Fluoroplastics, Melt-Processible Fluoropolymers, vol. 2, William Andrew, Inc, Norwich, NY, 2003, p. 547.
[7] B.J. Lyons, in: J. Scheirs (Ed.), In Modern Fluoropolymers, John Wiley and Sons, Ltd, Chichester, 1997, p. 340.
[8] S.V. Gangal, in: H.F. Mark, J.I. Kroschwitz (Eds.), Encyclopedia of Polymer Science and Engineering, vol. 16, John Wiley and Sons, New York, 1989, p. 597.
[9] C.H. Simon, Kaminski, Polym. Degrad. Stabil. 1998 (62) 1.
[10] German Patent DE 4334015 A1.
[11] I.J. Van der Walt, US Patent 6,797,913 to South African Energy Corporation September 2004.
[12] S. Ebnesajjad, V. Lishinsky, Machine Design, February 1999, p. 82.

# PART II

# Typical Properties of and Applications for Fluoropolymer Films

# 9 Polytetrafluoroethylene Films—Typical Properties and Applications

## 9.1 Skived Polytetrafluoroethylene Films

Polytetrafluoroethylene (PTFE) skived films are produced by skiving from molded sintered billets (see Section 3.1.1) and are stable up to 500°F (260°C) inert and unaffected by most of the chemicals with the exception of alkali metals. They are not sticky, easy to release, exhibit a low coefficient of friction, and have good dielectric properties. Skived PTFE films can be chemically surface treated and metalized when needed. Typical physical and mechanical properties of skived PTFE films are listed in Table 9.1 [1].

Typical applications for skived films include release and nonstick surfaces; chemical barriers; electrical insulations; low friction surfaces; food contact; heat-resistant adhesive tapes; gaskets; printed circuit boards; and chemical storage containers.

## 9.2 Cast Polytetrafluoroethylene Films

PTFE cast films are made up of very thin layers that have been individually deposited, sintered, and fused to produce a material having more uniform properties than can be achieved by any other manufacturing process (see Section 3.1.3). They are stable up to 500°F (260°C) and inert and unaffected by most of the chemicals with the exception of

**Table 9.1 Typical Properties of Polytetrafluoroethylene Skived Films [1].**

| Property | Value |
|---|---|
| Thickness, mil (μm) | 0.5–125 (12.5–3125) |
| Specific gravity | 2.14–2.19 |
| Tensile strength, minimum, psi (MPa) | 3500 (24.1) |
| Elongation, minimum (%) | 300 |
| Dielectric strength (V/mil) | 1680 |
| Hardness, Shore D | 54 |

**Applications of Fluoropolymer Films.** DOI: https://doi.org/10.1016/B978-0-12-816128-9.00009-X

**Table 9.2 Comparison of Cast and Skived Polytetrafluoroethylene Film Properties [2].**

| Film Type | Thickness (μm) | Tensile Strength (MPa) | | Break Elongation (%) | | Elastic Modulus (MPa) | |
|---|---|---|---|---|---|---|---|
| | | MD | TD | MD | TD | MD | TD |
| Skived | 76 | 52.3 | 40.4 | 450 | 360 | 469 | 517 |
| Cast | 68 | 35.8 | 34.5 | 530 | 510 | 434 | 434 |

*MD*, Machine direction; *TD*, transverse direction.

alkali metals. They are not sticky, easy to release, exhibit a low coefficient of friction, and have good dielectric properties. Besides pure PTFE versions, films consisting of individual layers of different polymers and polymer blends are available. In addition, colored, semiconductive, heat-weldable, and heat-bondable types are available. These films can be metalized and chemically treated for improving surface properties when needed. Typical physical and mechanical properties of cast PTFE films are listed in Table 9.2 [2].

Cast films are typically used as mold release films; vacuum bagging films; antistatic conductive films; as substrate for heat-resistant adhesive tapes; in electrical insulations; and printed circuit boards.

*Note*: When compared to skived films, cast films have two advantages, namely, absence of anisotropy (dependence of properties on direction) as shown in Table 9.3 [3] and possibility of producing

**Table 9.3** **Typical Properties of Polytetrafluoroethylene Cast Films [3].**

| Property | Test Method | Value |
|---|---|---|
| Tensile strength, MPa (psi) | ASTM D882 | 29.6 (4300) |
| Elongation at break (%) | ASTM D882 | 400 |
| Elastic modulus, MPa (psi) | ASTM D882 | 413 (60,000) |
| Continuous service temperature, °C (°F) | | 260 (500) |
| Dielectric constant | ASTM D150 | 2.0 |
| Dielectric strength (V/mil) | ASTM D149 | 4200 |

**Table 9.4** **Unsintered Polytetrafluoroethylene Films, Typical Properties [4].**

| | Value | |
|---|---|---|
| Property | English Units | Metric Units |
| Thickness | 2.0–3.0 mil | 0.051–0.076 mm |
| Specific gravity | 1.6 | 1.6 |
| Tensile strength, MD (minimum) | 1800 psi | 12.4 |
| Elongation at break (minimum) | 50 | 50 |

*MD*, Machine direction.

composite constructions from different fluoropolymers and/or colors as discussed in Sections 3.1.3 and 3.1.6.1.

## 9.3 Unsintered Polytetrafluoroethylene Films

Unsintered PTFE films are produced from PTFE fine powders by extrusion and subsequent calendering (see Section 3.1.2). The process introduces preferred orientation, which results in greater strength and lower elongation in the machine direction. Typical properties are listed in Table 9.4 [4].

## References

[1] CHEMFILM® T-100 Premium Skived Properties, <www.films.saint-gobain.com>, 2017.
[2] S. Ebnesajjad, Fluoroplastics, second ed., Non-Melt Processible Fluoropolymers, vol. 1, Elsevier, Oxford, 2014, p. 295.
[3] CHEMFILM® High Performance Films, Document AFF-3018-1016-SGCS, Saint Gobain, <www.films.saint-gobain.com>, 2016.
[4] FLUOROWRAP® E125 Extruded PTFE Film Properties, <www.films.saint-gobain.com>, 2017.

# 10 FEP Films, Typical Properties and Applications

FEP films are transparent; thermoplastic; and can be heat sealed, thermoformed, vacuum formed, heat bonded, welded, metalized, laminated, and combined with dozens of other materials. They can also be used as an excellent hot-melt adhesive. FEP films offer high chemical resistance; heat resistance; good dielectric properties; mechanical toughness, and long-time weatherability. More on this subject is discussed in Section 2.2.2.1. Typical properties of FEP films are summarized in Table 10.1.

**Table 10.1** **Typical Properties of FEP Films [1,2].**

| Property | Value |
|---|---|
| Tensile strength at break, MPa (psi) | 21 (3000) |
| Elongation at break (%) | 300 |
| Yield point, MPa (psi) | 12 (1700) |
| Tear strength, initial (Graves), N (g force) | 2.65 |
| Tear strength, propagation (Elmendorf), N (g) | 1.23 |
| Melt point, °C (°F) | 260–280 (500–536) |
| Dielectric strength (1 mm film), kV/mm (V/mil) | 260 (6500) |
| Dielectric constant, 25°C (77°F) | 2.0 |
| Dissipation factor, 25°C (77°F) | 0.0002–0.0007 |
| Surface resistivity (Ω) | $>10^{18}$ |
| Weatherability, continuous exposure in Florida, 2000 h | No adverse effect |
| Permeability, gas, $cm^3/m^2$ 24 h bar | |
| Carbon dioxide | $25.9 \times 10^3$ |
| Hydrogen | $34.1 \times 10^3$ |
| Nitrogen | $5.0 \times 10^3$ |
| Oxygen | $11.6 \times 10^3$ |
| Water vapor transmission, $g/m^2$ 24 h bar | 1 |
| Density, $kg/m^3$ ($lb/ft^3$) | 2150 (134) |
| Coefficient of friction, kinetic (film to steel) | 0.1–0.3 |
| Refractive index | 1.341–1.347 |
| Solar transmission | 96 |
| Critical surface tension (mN/m) | 17.8 |

**Applications of Fluoropolymer Films.** DOI: https://doi.org/10.1016/B978-0-12-816128-9.00010-6

Typical applications for FEP films are in solar collector windows, in pharmaceutical packaging; sealing of PTFE-coated fabrics; antigraffiti coverings; protection against extreme corrosion, fading or cracking; as mold release in aircraft industry; in printed circuit board manufacture; as autoclavable storage bags or liners; as lens covers, and in electret condenser microphone.

## References

[1] J.G. Drobny, in: J.G. Drobny (Ed.), Landolt-Börnstein, Specialty Thermoplastics, Advanced Materials and Technologies, vol. 13, Springer-Verlag, Heidelberg, 2015.

[2] Teflon™ FEP Fluoropolymer Film, Document C-10597(9/17), The Chemours Company, FC, LLC, 2017.

# 11 Perfluoroalkoxy Resin Films—Typical Properties

## 11.1 PFA Films

PFA films are highly chemically inert and solvent resistant; have low permeability to liquids, gases, moisture, and organic vapors; and exhibit very good dielectric properties, high continuous service temperature, favorable antistick and low frictional properties, as well as good weathering and optical properties. Additional features of PFA films are superior creep resistance at high temperatures, excellent low-temperature toughness, and exceptional flame resistance. Typical properties of PFA films are summarized in Table 11.1.

**Table 11.1 Typical Properties of PFA Films [1,2].**

| Property | Value |
|---|---|
| Tensile strength at break, MPa (psi) | 21 (3000) |
| Elongation at break (%) | 300 |
| Yield point, MPa (psi) | 12 (1700) |
| Tear strength, initial (Graves), N (g) | 4.90 (500) |
| Tear strength, propagation (Elmendorf), N (g) | 0.74 (75) |
| Melt point, °C (°F) | 302–310 (575–590) |
| Dielectric strength (1 mm film), kV/mm (V/mil) | 260 (6500) |
| Dielectric constant, 25°C (77°F) | 2.0 |
| Dissipation factor, 25°C (77°F) | 0.0002–0.0007 |
| Surface resistivity (Ω) | $>10^{17}$ |
| Permeability, gas, $cm^3/m^2$ 24 h bar | |
| Carbon dioxide | $14 \times 10^3$ |
| Nitrogen | $2.0 \times 10^3$ |
| Oxygen | $14 \times 10^3$ |
| Water vapor transmission, $g/m^2$ 24 h bar | 8 |
| Density, $kg/m^3$ ($lb/ft^3$) | 2150 (134) |
| Coefficient of friction, kinetic (film to steel) | 0.1–0.3 |
| Refractive index | 1.350 |
| Solar transmission (%) | 95 |
| Critical surface tension (mN/m) | 22.0 |

**Applications of Fluoropolymer Films.** DOI: https://doi.org/10.1016/B978-0-12-816128-9.00011-8

Typical applications of PFA films are in chemical processing, wire and cable, electronics, mold release in processing composites, circuit board laminates, chemical pumps, tank linings, shrink roll covers, and rupture discs and gaskets.

## 11.2 MFA Films

MFA films offer a combination of excellent dielectric properties across a wide temperature and frequency ranges; chemical and stress-crack resistance similar to PFA, a continuous service temperature of 230°C (446°F), and the highest clarity of any fluoropolymer film [1]. Other properties of MFA films are outstanding flex life, stress-crack resistance, favorable antistick properties, and very good weatherability. MFA film is a clear, transparent product that can be heat sealed, thermoformed, welded, metallized, or laminated to a wide variety of materials. Typical properties of MFA films are summarized in Table 11.2.

Examples of applications of MFA films are found in chemical process industry (tank linings and roll covers), in solar collectors,

**Table 11.2 Typical Properties of MFA Films [3].**

| Property | Value |
|---|---|
| Tensile strength at break, MPa (psi) | 21 (3000) |
| Elongation at break (%) | 300 |
| Yield point, MPa (psi) | 24.1–30.2 (3500–4375) |
| Melt point, °C (°F) | 280–290 (536–554) |
| Maximum service temperature, °C (°F) | 230 (446) |
| Water absorption (%) | $<0.03$ |
| Dielectric strength (1 mm film), kV/mm (V/mil) | 197 (5000) |
| Dielectric constant, 25°C (77°F), 1 MHz | 2.0 |
| Dissipation factor, 25°C (77°F), 1 MHz | 0.0005 |
| Surface resistivity (Ω) | $>10^{15}$ |
| Oxygen index (%) | 95 |
| Density, kg/m$^3$ (lb/ft$^3$) | 2150 (132) |
| Coefficient of friction, kinetic (film to steel) | 0.2 |
| Refractive index | 1.35 |
| Solar transmission (%) | $>90$ |
| Critical surface tension (mN/m) | 22.0 |

UV-protective films, release films in composite manufacture, in circuit boards, and as electrical insulations.

It should be noted that the melting temperature of MFA resin is 20°C (36°F) lower than that of the PFA resin [2,3].

## References

[1] M. Pozzoli, G. Vita, V. Arcella, in: U.K. Chichester (Ed.), Modern Fluoropolymers, Scheirs, John Wiley & Sons, Ltd, 1997, p. 386.

[2] J.G. Drobny, Specialty Thermoplastics, Landolt-Börnstein, Advanced Thermoplastics, vol. 13, Springer Verlag, Heidelberg, Germany, 2015.

[3] Saint Gobain Norton® MFA Fluoropolymer Film, MatWeb, <www.matweb.com>, 2019.

# 12 ETFE Films—Typical Properties and Applications

ETFE films are tough, transparent, and thermoplastic films with high resistance to impact and tearing, outstanding antistick, and low frictional properties. They are chemically inert and solvent resistant to virtually all chemicals, except molten alkali metals, gaseous fluorine, and certain complex halogenated compounds. They also have a relatively low permeability to liquids, gases, moisture, and organic vapors; are inert to outdoor exposure; and have very good dielectric properties and wide continuous service temperature range. ETFE films can be heat sealed, thermoformed, vacuum formed, heat bonded, welded, metalized, laminated (combined with dozens of other materials), and used as an

**Table 12.1 Typical Properties of ETFE Films [1,2].**

| Property | Value |
|---|---|
| Tensile strength at break, N/mm$^2$ (psi) | 41 (6000) |
| Elongation at break (%) | 300 |
| Tear strength, initial (Graves), N (g) | 4.90 (500) |
| Tear strength, propagation (Elmendorf), N (g) | 0.74 (75) |
| Melt point, °C (°F) | 260–280 (500–536) |
| Dielectric strength (1 mm film), kV/mm (V/mil) | 160 (4000) |
| Dielectric constant, 25°C (77°F) | 2.6 |
| Volume resistivity, at 170°C (338°F), Ω cm | $>1\times10^{17}$ |
| Dissipation factor, 25°C (77°F) | 0.0007 |
| Permeability, gas, cm$^3$/m$^2$ 24 h atm | |
| Carbon dioxide | $3.9\times10^3$ |
| Nitrogen | $0.5\times10^3$ |
| Oxygen | $1.6\times10^3$ |
| Water vapor transmission, g/m$^2$ 24 h bar | 7.8 |
| Density (kg/m$^3$) | 1700 (106) |
| Coefficient of friction, kinetic (film to steel) | 0.2–0.3 |
| Refractive index @25°C | 1.42 |
| Solar transmission (%) | 90 |
| Critical surface tension (mN/m) | 16.5 |

**Applications of Fluoropolymer Films.** DOI: https://doi.org/10.1016/B978-0-12-816128-9.00012-X

excellent hot-melt adhesive. Typical properties of ETFE films are summarized in Table 12.1.

Examples of applications of ETFE films are anticorrosive linings; composite part mold release film; industrial roll covers; circuitry; pharmaceutical cap liners; sterile packaging; cable insulation; hot-melt adhesive; microphone electret membranes; photovoltaic cell glazing (back sheet); transparent or translucent ETFE film constructions, in the form of roof and wall structures made from segmented air cushions; and hot-melt adhesives.

## References

[1] J.G. Drobny, Specialty Thermoplastics, Landolt-Börnstein, Advanced Thermoplastics, vol. 13, Springer Verlag, Heidelberg, Germany, 2015.

[2] Tefzel™ ETFE Fluoropolymer Film, Properties Bulletin, The Chemours Company, <www.chemours.com>, 2019.

# 13 PCTFE Films—Typical Properties and Applications

PCTFE films are tough transparent thermoplastic films exhibiting the lowest permeation rates of any thermoplastic. They exhibit superior clarity (their low, only 1% haze allows these films to disappear as overlays and liners). They are chemically inert, resisting to most chemicals and solvents and flame resistant and possess a very good formability, which makes them suitable for complex thermoforming. More on this subject is in Section 2.2.4.4. Typical properties for PCTFE films are shown in Table 13.1.

**Table 13.1** **Typical Properties of PCTFE Films.**

| Property | Value |
|---|---|
| Tensile strength at break, MPa (psi) | |
| MD | 21–32 (3000–5000) |
| TD | 21–28 (3000–4000) |
| Elongation at break (%) | |
| MD | 50–125 |
| TD | 50–125 |
| Tear strength, initial (Graves) (g/mil) | |
| MD | 425–525 |
| TD | 425–525 |
| Melt point, °C (°F) | 206 (403) |
| Continuous service temperature, °C (°F) | 180 (356) |
| Dielectric strength (1 mm film) (kV/mm) | 55–81 |
| Dielectric constant, 25°C (77°F) at $10^3$ Hz | 2.3 |
| Dissipation factor, 25°C (77°F) at $10^3$ Hz | 0.0002 |
| Permeability, gas, $cm^3/m^2$ 24 h bar | |
| Carbon dioxide | 150 |
| Hydrogen | – |
| Nitrogen | 10 |
| Oxygen | 60 |

**Applications of Fluoropolymer Films.** DOI: https://doi.org/10.1016/B978-0-12-816128-9.00013-1

**Table 13.1 Typical Properties of PCTFE Films.—cont'd**

| Property | Value |
|---|---|
| Water vapor transmission rate, $g/m^2/24$ h bar (g/100 in.$^2$ 24 h bar) | 0.05 (0.003) |
| Specific gravity | 2.12−2.16 |
| Oxygen index | 100 |
| Static coefficient of friction (film to steel) | 0.26−0.42 |
| Refractive index | 1.435 |
| Solar transmission (%) | >90 |
| Critical surface tension (mN/m) | 30.8 |

*MD*, Machine direction; *TD*, transverse direction.

Examples of applications of PCTFE films are electroluminescent panel overlays, blister packaging, and encapsulation of electronics.

# Further Reading

J.G. Drobny, Specialty Thermoplastics, Landolt-Börnstein, vol. 13, Springer Verlag, Heidelberg, Germany, 2015.

Honeywell International, Inc., Honeywell Aclar Rx20e, Honeywell International, Inc., <www.honeywell.com>, 2010.

# 14 ECTFE Films—Typical Properties and Applications

ECTFE films offer the combination of heat stability, electrical characteristics, and barrier properties. They exhibit the highest dielectric strength of all fluoropolymer films and have outstanding resistance to weathering and high energy radiation [1,2]. They also provide the highest abrasion resistance of any fluoropolymer film available, exceptional chemical resistance as well as very high tensile strength, and flexural modulus [3,4]. Typical properties for ECTFE films are listed in Table 14.1.

ECTFE films are suited for use in sensitive electrical applications, such as for electrical tapes, cable insulation, printed circuits, capacitors, chlorine cells, flat cable constructions, and solar collectors, in caustic,

**Table 14.1** **Typical Properties of ECTFE Films [1–4].**

| Property | Value |
|---|---|
| Tensile strength at break, MPa (psi) | 45 (6500) |
| Elongation at break (%) | 250 |
| Yield point, MPa (psi) | 30–32 |
| Hardness, Shore D | 75 |
| Melt point, °C (°F) | 240 (465) |
| Continuous use temperature, °C (°F) | 165 (330) |
| Dielectric strength (1 mil film), V/mil (kV/mm) | 5500 (220) |
| Dielectric constant, 21°C (70°F) 1 kHz | 2.6 |
| Dissipation factor, 21°C (70°F) | <0.05 |
| Flammability rating, UL94 | V-0 |
| Critical surface tension (mN/m) | 32 |
| Specific gravity | 1.68 |
| Coefficient of friction, static/dynamic versus itself | 0.2 |
| Refractive index @ 25°C | 1.447 |
| Haze (%) | 3–5 |
| Water absorption, 24 h @ 23°C, % | <0.01 |
| Taber abrasion, CS-17 wheel, 500 g load, mg/1000 rev | 5.00 |

**Applications of Fluoropolymer Films.** DOI: https://doi.org/10.1016/B978-0-12-816128-9.00014-3

high temperature environments. Other applications include aircraft cabin interiors and fuel cell membranes.

## References

[1] TCI ECTFE Extruded Films, Properties, <www.textilescoated.com>, 2019.

[2] CHEMFILM®ECTFE Film, Typical Physical Properties, AFF-1005R-1116-SGCS, Saint-Gobain Performance Plastics, 2016.

[3] CS Hyde Company, Halar®500C Film, Product Information, <www.cshyde.com>, 2019.

[4] J.G. Drobny, Specialty thermoplastics, Landolt-Börnstein, vol. 13, Springer Verlag, Heidelberg, 2015.

# 15 PVDF Films—Typical Properties and Applications

PVDF films exhibit excellent physical and mechanical properties of a polymer: dielectric strength, chemical resistance, very good transmittance of solar energy, high heat resistance, high resistance to creep and fatigue, are stable to UV radiation and effects of weather, generate small amount of smoke, and are FDA compliant. Under certain

**Table 15.1 Typical Properties of PVDF Film [1–3].**

| Property | Value |
|---|---|
| Tensile strength at break, MPa (psi) | 35–48 (6000–7000) |
| Elongation at break (%) | 50–250 |
| Yield point, MPa (psi) | 48 (7000) |
| Tear strength, propagation (Elmendorf), (g/mil) | 100 |
| Hardness, Shore D | 65–82 |
| Melt point, °C (°F) | 170 (338) |
| Continuous use temperature (°C) | 150 (302) |
| Embrittlement temperature (°C) | −60 |
| Dielectric strength (1 mm film), kV/mm (V/mil) | 10–27 (1.900) |
| Dielectric constant, 25°C (77°F), 1 kHz | 7.5–9 |
| Dissipation factor, 25°C (77°F), 1 kHz | 0.01–0.03 |
| Surface resistivity (Ω) | $\geq 10^{14}$ |
| Permeability, gas, $cm^3/m^2$ 24 h bar | |
| Carbon dioxide | 100 |
| Air | 7 |
| Nitrogen | 30 |
| Oxygen | 20 |
| Water vapor transmission, 100 μm film, $g/m^2$ 24 h bar | 2 |
| Density ($kg/m^3$) | 1700–1800 |
| Coefficient of friction, static (film to polished steel) | 0.14–0.17 |
| Abrasion resistance, Taber CS, 1 kg, mg/1000 cycles | 5–10 |
| Refractive index | 1.40–1.42 |
| Solar transmission (%) | 90 |
| Oxygen index (%) | 43 |
| Water absorption, 24 h @ 23°C (73°F), % | 0.03 |
| Critical surface tension (mN/m) | 25 |

**Applications of Fluoropolymer Films.** DOI: https://doi.org/10.1016/B978-0-12-816128-9.00015-5

conditions, PVDF films become piezoelectric and pyroelectric (see Section 2.2.4.6). Typical properties of PVDF films are shown in Table 15.1.

Examples of applications of PVDF films are: filters, diaphragms, release films, chemical-resistant tank linings, fuel cell seals, and medical bags. Their piezoelectric properties are utilized in soundproof telephone headset, infrared sensing, a respiration monitor, a high-fidelity electric violin, hydrophones, keyboards, and printers [1–3].

## References

[1] CS Hyde Company, Kynar 740 Resin PVDF Film, Product Information, <www.cshyde.com>, 2019.

[2] TCI PVDF Fluoropolymer Extruded Films, Properties, <www.textilcoated.com>, 2019.

[3] J.G. Drobny, Specialty Thermoplastics, Landolt-Börnstein, vol. 13, Springer Verlag, Heidelberg, 2015.

# 16 THV Films—Typical Properties

THV films provide a combination of performance advantages, such as low processing temperatures, ability to bond to elastomers and hydrocarbon-based plastics, flexibility, and optical clarity. Typical properties are shown in Table 16.1.

Examples of applications are automotive, chemical processing, multilayer tubes and hoses, semiconductors, solar energy, optical fiber, and architectural and protective coatings, which are broadly discussed in Section 2.2.4.7.

**Table 16.1** **Typical Properties of THV Films [1].**

| Property | Value |
|---|---|
| Tensile strength at break, MPa (psi)[a] | 20–29 |
| Elongation at break (%)[a] | 420–600 |
| Melt point, °C (°F)[a] | 120–225 |
| Maximum service temperature, short term, °C (°F)[a] | 150 |
| Maximum service temperature, long term (20,000 h), °C (°F) | 80–120 |
| Embrittlement temperature, °C (°F)[a] | −75 to −85 |
| Hardness, Shore D[a] | 44–58 |
| Dielectric strength (0.25 mm film), kV/mm[a] | 48–56 |
| Dielectric constant, 25°C (77°F), $10^6$ Hz[a] | 4.7–5.7 |
| Dissipation factor, 25°C (77°F), $10^6$ Hz[a] | 0.09–0.14 |
| Permeability, film 0.1 mm thick, 20°C, gas, $cm^3/m^2$ 24 h atm | |
| Carbon dioxide | 2060 |
| Air | 291 |
| Nitrogen | 217 |
| Oxygen | 696 |
| Water vapor transmission, $g/m^2$ 24 h bar | 1.73 |
| Specific gravity[a] | 1.95–2.06 |
| Coefficient of friction, kinetic (film to steel) | |
| Refractive index @ 25°C[a] | 1.350–1.363 |

[a]Depends on composition.

**Applications of Fluoropolymer Films.** DOI: https://doi.org/10.1016/B978-0-12-816128-9.00016-7

## Reference

[1] J.G. Drobny, Specialty thermoplastics, Landolt-Börnstein, vol. 13, Springer Verlag, Heidelberg, Germany, 2015.

# 17 Teflon AF Films—Typical Properties

Teflon AF films can be produced from the resins by melt extrusion or from solutions by spin coating, spraying, and dipping techniques They exhibit high thermal, chemical, mechanical, and electrical stability; high permanent gas permeability; high creep resistance; low thermal conductivity; optical transparency over a wide wavelength range; lowest dielectric constant of any known solid plastic; and lowest index of refraction of any known plastic. Typical properties of the resins used for the films are summarized in Table 17.1.

**Table 17.1** **Typical Properties of Teflon AF (1600 and 2400, Respectively) [1].**

| Property | Value |
|---|---|
| Tensile strength at break, MPa (psi) | 26−27 |
| Elongation at break (%) | 8−17 |
| Yield point, MPa (psi) | 26−27 |
| Durometer, Shore D | 75−77 |
| Dielectric strength (0.11 mm film), kV/0.1 mm | 1.9.−2.1 |
| Dielectric constant, 25°C (77°F) | 1.90−1.93 |
| Dissipation factor, 25°C (77°F) | 0.0001−0.0003 |
| Taber abrasion, $cm^3$/2000 cycles | 0.107−0.2 |
| Permeability, gas, Barrer | |
| $H_2O$ (vapor) | 1142−4026 |
| Oxygen | 340−990 |
| Nitrogen | 130−490 |
| Carbon dioxide | 2800 |
| Specific gravity | 1.67−1.78 |
| Coefficient of friction, kinetic (film to steel) | |
| Refractive index | 1.29−1.31 |
| Optical transmission (%) | >95 |
| Water absorption (%) | <0.1 |
| Critical surface energy, dyn/cm (mN/m) | 15.6−15.7 |
| Contact angle with water (degrees) | 104−105 |

**Applications of Fluoropolymer Films.** DOI: https://doi.org/10.1016/B978-0-12-816128-9.00017-9

Examples of applications are deep UV-resistant films and components, interlayer dielectric, electret films, nanocomposite films, by spin coating or dip coating techniques, etc.

## Reference

[1] Teflon™ Amorphous Fluoroplastic Resins, Product Information, C-140233 (9/16), <www.nemours.com>, 2016.

# 18 DAI-EL T-530 Films—Typical Properties

DAI-EL T-530 films are transparent thermoplastic films with elastic properties and low extractables, excellent physical properties, and chemical resistance. Their properties can be enhanced by irradiation with electron beam. Typical properties of these films are shown in Table 18.1.

Applications for DAI-EL T-530 films are mainly in contact with chemicals, food, and pharmaceuticals.

**Table 18.1** **Typical Properties of DAI-EL T-530 Films [1].**

| Property | Value |
|---|---|
| Tensile strength, MPa (psi) | 12 |
| Elongation at break (%) | 650 |
| Durometer, JIS A | 67 |
| Tear strength (kN/m) | 28 |
| Taber abrasion wheel CS-17, 1000 g load, mg/1000 rev | 2 |
| Melt point, °C (°F) | 220 (428) |
| Oxygen index (%) | 66 |
| Dielectric strength (kV/mm) | 14 |
| Dielectric constant, 23°C $10^3$ Hz | 6.6 |
| Permeability, gas, $cm^3$ $mm/m^2$ 24 h atm | |
| Carbon dioxide | 111 |
| Nitrogen | 82 |
| Oxygen | 136 |
| Helium | 1715 |
| Specific gravity, 25°C | 1.89 |
| Coefficient of friction | 0.6 |
| Refractive index, $n_D^{20}$ | 1.357 |
| Total luminous transmittance (1 mm) (%) | 87 |
| Critical surface tension (mN/m) | 20.5 |

**Applications of Fluoropolymer Films.** DOI: https://doi.org/10.1016/B978-0-12-816128-9.00018-0

## Reference

[1] M. Tatemoto, T. Shimizu, in: J. Scheirs (Ed.), Modern Fluoropolymers, John Wiley & Sons Ltd., Chichester, 1997, p. 569.

# 19 Polyvinyl Fluoride Films—Typical Properties and Applications

Polyvinyl fluoride (PVF) is a semicrystalline, transparent-to-opaque thermoplastic polymer. Films produced from it have the following basic attributes: resistance to UV, weathering, wide variety of chemicals, solvents, and staining agents; good mechanical properties; strength and durability; good release and dielectric properties; piezoelectricity and pyroelectricity; and good moisture barriers [1]. These are widely discussed in Section 2.3.2.1.

PVF films are essentially available in two forms: oriented and unoriented. The properties of the two forms differ considerably, because orientation improves mainly the physical properties of the films [2]. The applicable processes are discussed in some detail in Chapter 5, Films

**Table 19.1 Typical Properties of Oriented Grade, 1 mil Thick Clear PVF Films [1,2].**

| Property | Value |
|---|---|
| Strength, min. (MD), MPa (psi) | 90 (13,050) |
| Ultimate elongation at break (%) | 95 |
| Tensile modulus (MD), MPa (kpsi) | 2075 (301) |
| Tear strength, propagating (MD), Elmendorf, kN/m (g/mil) | 7.4 (19.2) |
| Tear strength, initial (MD), Graves, kN/m (g/mil) | 163 (423) |
| Tear strength, initial (TD), Graves, kN/m (g/mil) | 185 (478) |
| Bursting strength, Mullen (MPa/m) | 13,057 (48.1) |
| Linear coefficient of expansion (MD) (m/m K) | $8.8 \times 10^{-6}$ |
| Linear coefficient of expansion (TD) (m/m K) | $7.1 \times 10^{-5}$ |
| Shrinkage, max (TD), % at °C | 5 at 170 |
| Specific heat, kJ/kg K (cal/g °C) | 1.01 (0.24) |
| Aging in air, hours to embrittlement | 3000 |
| Moisture absorption (%) | $<0.5$ |
| Specific gravity | 1.39 |
| Coefficient of friction, kinetic (film to metal) | 0.21 |
| Refractive index | 1.46 |
| Solar transmission (%) | $>90$ |
| Critical surface tension (mN/m) | 28.0 |

**Applications of Fluoropolymer Films.** DOI: https://doi.org/10.1016/B978-0-12-816128-9.00019-2

**Table 19.2 Typical Properties of Unoriented Grade, 1 mil Thick Clear PVF Films [2,3].**

| Property | Value |
|---|---|
| Tensile strength at break, MPa (psi) | 41 (6000) |
| Ultimate elongation (%) | 200 |
| Tear strength, initial (MD), Graves, kN/m (g/mil) | 550 (212) |
| Tear strength, initial (TD), Graves, kN/m (g/mil) | 550 (212) |
| Moisture absorption (%) | 0.5 |
| Temperature range, °C (°F) | |
| Continuous use | −72 to 107 (−98 to 225) |
| Short cycle | Up to 170 (338) |
| Linear coefficient of expansion (MD) (m/m K) | $9 \times 10^{-5}$ |
| Linear coefficient of expansion (TD) (m/m K) | $9 \times 10^{-5}$ |
| Shrinkage, maximum, % at 170°C | 2 |
| Specific heat, kJ/kg K (cal/g °C) | 1.01 (0.24) |
| Dielectric strength (1 mil film) (V/mil) | 3000 |
| Dielectric constant | 7 |
| Dissipation factor (%) | 0.2 |
| Volume resistivity (Ω/cm) | $4 \times 10^{13}$ |
| Critical surface tension (mN/m) | 32 |
| Moisture vapor transmission, g/m$^2$ day atm | 30 |
| Coefficient of friction, kinetic (film to steel) | 0.21 |
| Refractive index | 1.46 |
| Haze internal (%) | 0.6 |

from polyvinyl fluoride. Typical properties of the two types of PVF films are summarized in Tables 19.1 and 19.2.

Examples of the applications of PVF films are as surface protecting laminates in the aircraft and architectural industry to improve the chemical and UV resistance and to provide an easy-to-clean surface. Important architectural PVF film applications include wall coverings, residential and commercial roofing, siding, air-inflated structures, canopies, awnings, and stadium domes. They are also used as release films for the manufacture of fiber-reinforced composites and for transfer printing and as a backsheet material for solar panels. PVF films often require a surface treatment to improve bonding to other materials [3]. Among these are flame treatment, electric discharge, chemical etching, and plasma treatment (see Section 2.3.3.1). Ref. [4] will provide additional data.

## References

[1] S. Ebnesajjad, Chapter 7 Polyvinyl Fluoride—Technology and Applications of PVF, Elsevier, Oxford, 2013.
[2] S. Ebnesajjad, Chapter 6 Polyvinyl Fluoride—Technology and Applications of PVF, Elsevier, Oxford, 2013.
[3] S. Ebnesajjad, Polyvinyl Fluoride—Technology and Applications of PVF, Elsevier, Oxford, 2013, p. 164.
[4] J.G. Drobny, Specialty Thermoplastics, Landolt-Börnstein, Advanced Thermoplastics, vol. 13, Springer Verlag, Heidelberg, 2015.

# PART III

# Commercial Grades of Fluoropolymer Films and Their Applications

# 20 Commercial Grades of Fluoropolymer Films

In this chapter, grades and properties of some of the selected commercial fluoropolymer products are presented. The sources are the data published by individual companies. It should be noted that not all companies producing and supplying fluoropolymer films use data in a uniform fashion, for example, often different units are shown. Where possible, units have been changed for the sake of uniformity. In addition, sometimes the number of different grades is very large, so a few typical ones have been selected. For additional information the reader is advised to study the entire range of products listed on the website of a given company. Websites of the major manufacturers, suppliers, and distributors are summarized in Appendix 1. Our intention is to provide much useful information for the reader, but it in no way guarantee accuracy and claim the data specifications. The reader is advised to contact the manufacturer, supplier and/or distributor for that.

## 20.1 Polytetrafluoroethylene Films

### 20.1.1 Skived Films

Skived polytetrafluoroethylene (PTFE) films are produced by compression molding, sintering, and skiving as discussed in Section 3.1.1. These films provide a low-friction, nonstick solution for a number of high-temperature and high-dielectric applications. They are supplied in a roll form, sheets, and tapes and etchable on one or both sides. The grades available are made from homopolymer PTFE, modified PTFE (heat weldable), and carbon black–filled antistatic conductive PTFE compounds. The data pertaining to skived PTFE films are presented in Tables 20.1–20.7.

### 20.1.2 Cast Films

Commercial cast PTFE-based films are made up of very thin layers that have been individually deposited, sintered, and fused together to produce a material having more uniform properties than can be achieved by any other manufacturing process. More details regarding

**Applications of Fluoropolymer Films.** DOI: https://doi.org/10.1016/B978-0-12-816128-9.00020-9

**Table 20.1 CHEMFILM T-100 Premium Polytetrafluoroethylene Skived Film, Properties.**

| Property | Data |
|---|---|
| Thickness, mil | 0.5–125 |
| Density, g/cm$^3$ | 2.14–2.19 |
| Tensile strength, minimum, psi | 3500 |
| Elongation at break, minimum, % | 300 |
| Dielectric strength, V/mil | 1680 |
| Hardness, Shore D | 54 |

*Source*: Data from <www.filmssaint-gobain.com>, 2017.

**Table 20.2 CHEMFILM T-500 High Modulus Polytetrafluoroethylene Skived Film, Properties.**

| Property | T-500 | | | |
|---|---|---|---|---|
| Thickness, mil | 2.0 | 3.0 | 4.0 | 5.0 |
| Thickness tolerance, ± mil | 0.3 | 0.3 | 0.3 | 0.3 |
| Tensile strength, minimum, psi | 10,000 | 10,000 | 10,000 | 8000 |
| Elongation at break, minimum, % | 100 | 100 | 125 | 125 |
| Dielectric strength, total volts | 8000 | 9000 | 13,000 | 13,000 |
| Specific gravity | 2.14 | | | |

*Source*: Data from <www.filmssaint-gobain.com>, 2017.

**Table 20.3 Skived Polytetrafluoroethylene Film DW 2000, Properties.**

| Property | ASTM Test Method | Data |
|---|---|---|
| Tensile strength, psi (MPa) | D882 | 3000 (20.68) |
| Elongation at break, % | D882 | 325 |
| Dielectric strength (film 1 mil thick), V | D149 | 2000 |
| Specific gravity | D792 | 2.14–2.19 |
| Maximum operating temperature, °F (°C) | 500°F (260°C) | |

*Source*: Data from <www.dewal.com>, 2017.

**Table 20.4** **DW 200 Skived Polytetrafluoroethylene (PTFE) Film, Properties.**

| Property | ASTM Test Method | Data |
|---|---|---|
| Tensile strength, psi (MPa), MD[a] | D882 | 6000 (41.37) |
| Elongation at break, % | D882 | 500 |
| Dielectric strength (film 2 mil thick), V | D149 | 2800 |
| Maximum operating temperature, °F (°C) | | 500°F (260°C) |

*Note*: Based on modified PTFE.
[a]Measured in machine direction.
*Source*: Data from <www.dewal.com>, 2017.

**Table 20.5** **DW 2012AE Skived Modified Polytetrafluoroethylene (PTFE) Film, Properties.**

| Property | ASTM Test Method | Data |
|---|---|---|
| Tensile strength, psi (MPa) | D882 | 6000 (41.37) |
| Elongation at break, % | D882 | 500 |
| Dielectric strength, V/mil | D149 | 2800 |
| Maximum operating temperature, °F (°C) | | 500°F (260°C) |

*Note*: Based on modified PTFE, etched for adhesive bonding.
*Source*: Data from <www.dewal.com>, 2017.

**Table 20.6** **DW 103 Skived Conductive Polytetrafluoroethylene Film, Properties.**

| Property | ASTM Test Method | Data |
|---|---|---|
| Tensile strength, psi (MPa) | D882 | 2000 (13.79) |
| Elongation at break, % | D882 | 0.55 |
| Electrical resistance for 0.250″ wide at 3.5 mil thick | – | 35k Ω/ft |
| Electrical resistance for 0.500″ wide at 3.5 mil thick | – | 35k Ω/ft |
| Electrical resistance for 0.750″ wide at 3.5 mil thick | – | 10.5k Ω/ft |
| Maximum operating temperature, °F (°C) | | 500°F (260°C) |

*Source*: Data from <www.dewal.com>, 2017.

**Table 20.7 CHEMFILM 0200 Skived Modified Polytetrafluoroethylene Film, Properties.**

| Property | Data |
|---|---|
| Thickness, mil (mm) | 0.5−120 (0.0125−3.0) |
| Typical yield for 0.1 mm film, $m^2$/kg | 0.23 |
| Tensile strength, minimum, N/$mm^2$ (psi) | 25 (3626) |
| Elongation at break, minimum, % | 400 |
| Temperature, °C (°F) | −200 to +260 (−328 to +500) |

*Note*: Better cold flow and flex life behavior, lower permeability. The manufacturing site complies with ISO 9001 and IATF6949. The product conforms to FDA 21 CFR 177.1550.
*Source*: Data from <www.filmssaint-gobain.com>, 2018.

this process are in Section 3.1.3. Pigments and other additives may be added to individual layers to customize performance. Cast PTFE films can be produced in a clear and filled form, and in a wide array of colors and as a multilayer combination of fluoropolymers and their blends. They are marketed as *mold release films*, *vacuum bagging films*, *monolayer films*, *cementable adhesive coated films*, *antistatic conductive films*, and *heat weldable films*. As of this writing, there are two manufacturers producing and offering cast PTFE films. The data pertaining to cast PTFE films are in Tables 20.8−20.13.

### 20.1.3 Unsintered Polytetrafluoroethylene Films

Unsintered PTFE films are produced from PTFE *fine powders* by paste extrusion followed by calendering as described in detail in Section 3.1.2. The polymer is fibrillated during paste extrusion, imparting good strength and low elongation in the machine direction. Commercial products are available from several manufacturers and suppliers in a variety of grades with different properties and colors, as shown in Tables 20.14−20.16.

## 20.2 Fluorinated Ethylene Propylene Films

Fluorinated ethylene propylene films, produced from copolymers of hexafluoropropylene (HFP) and tetrafluoroethylene (TFE), are transparent; thermoplastic films can be heat sealed, thermoformed, vacuum

**Table 20.8** **CHEMFILM MR, Cast Film, Properties.**

| Property | ASTM Test Method | Units | Values |
|---|---|---|---|
| Tensile strength | D882 | MPa (psi) | 31.0 (4500) |
| Elongation at break | D882 | % | 400 |
| Elastic modulus | D882 | MPa (psi) | 379 (55,000) |
| Yield stress | D882 | MPa (psi) | 9.65 (1400) |
| Resistance to tear initiation | D1004 | MPa (psi) | 3.4 (500) |
| Propagation tear strength | D1938 | kgF (lb) | 0.045 (0.10) |
| Bursting strength | D774 | MPa (psi) | 0.51 (75) |
| Dimensional stability, shrinkage, % | 30 min 500°F | – | 3% max |
| Coefficient of linear thermal expansion | D696 | in./in. °F | $5.5 \times 10^{-5}$ |
| Continuous service temperature | − 188°C to +260°C (−400°F to +500°F) | | |

*Note*: Pure PTFE, available also in a wide array of colors. *PTFE*, Polytetrafluoroethylene.
*Source*: Data from Saint Gobain, AFF-3017-1016 SGCS, 2016; <www.plastics-saint-gabain.com>.

**Table 20.9** **CHEMFILM VB, Cast Film, Properties.**

| Property | ASTM Test Method | Units | Values |
|---|---|---|---|
| Gauge | D882 | mil | 3.0 |
| Tensile strength | D882 | MPa (psi) | 29.6 (4300) |
| Elongation at break | D882 | % | 550 |
| Elastic modulus | D882 | MPa (psi) | 379 (55,000) |
| Yield stress | D882 | MPa (psi) | 9.65 (1400) |
| Yield strain | D1004 | kPa (psi) | 17.2 (2.5) |
| Coefficient of linear thermal expansion | D696 | in./in. °F | $5.5 \times 10^{-5}$ |
| Maximum continuous service temperature | D696 | 260 (500) | °C (°F) |

*Note*: Pure multilayer PTFE cast film, clear, for vacuum bagging. *PTFE*, Polytetrafluoroethylene.
*Source*: Data from Saint Gobain, AFF-3022-1016-SGCS; <www.plastics-saint-gabain.com>.

**Table 20.10 CHEMFILM DF100 and Type C/CD Monolayer Cast Films, Properties.**

| Property | ASTM Test Method | Units | Value |
|---|---|---|---|
| **Mechanical** | | | |
| Tensile strength | D882 | MPa (psi) | 29.6 (4300) |
| Elongation at break | D882 | % | 400 |
| Elastic modulus | D882 | MPa (psi) | 413 (60,000) |
| Continuous service temperature | | °C (°F) | −188 to 260 (−400 to 500) |
| **Electrical** | | | |
| Dielectric constant, 60–$10^8$ Hz | D150 | | 2.0 |
| Power/dissipation | D150 | – | 0.0001 |
| Surface resistivity | D257 | Ω | $9 \times 10^7$ |
| Volume resistivity | D257 | Ω cm | $>10^{15}$ |
| Dielectric strength | D149 | V/mil | 4200 |
| Surface arc resistivity | D495 | | Does not arc |

*Note*: CHEMFILM DF100 and Type C and CD (duplex wound) are thin multilayer cast PTFE films. Combining the uniform tensile properties, excellent dielectric properties and isotropic shrinkage with temperature, they act as highly suitable films for dielectric insulation. *PTFE*, Polytetrafluoroethylene.
*Source*: Data from Saint Gobain, AFF-3018-1016 SGCS, 2016; <www.plastics-saint-gobain.com>.

**Table 20.11 CHEMFILM Types DF1100, DF1200 Cast Films, Properties.**

| Property | ASTM Test Method | Units | Value |
|---|---|---|---|
| Tensile strength | D882 | MPa (psi) | 29.6 (4300) |
| Elongation at break | D882 | % | 400 |
| Elastic modulus | D882 | MPa (psi) | 413 (60,000) |
| Dielectric constant | D150 | 60–$10^8$ Hz | 2.0 |
| Power/dissipation | D150 | – | $<0.0001$ |
| Surface resistivity | D257 | – | $9 \times 10^7$ |
| Volume resistivity | D257 | – | $>10^{15}$ |
| Dielectric strength | D149 | V/mil | 4200 |
| Surface arc resistivity | D495 | | Does not arc |
| Continuous service temperature, °C (°F) | **DF1100**: −188 to 260 (−400 to 500)<br>**DF1200**: −90 to 140 (−100 to 300) | | |

*Note*: CHEMFILM DF1200 is DF1100 laminated to 2 mil of acrylic pressure sensitive adhesive.
*Source*: Data from Saint Gobain, AFF-3019-1016 SGCS, 2016; <www.plastics-saint-gobain.com>.

**Table 20.12** **CHEMFILM DF1400, DF1471, Cast Films, Properties.**

| Property | ASTM Test Method | Units | Value |
|---|---|---|---|
| Tensile strength | D882 | MPa (psi) | 27.6 (4000) |
| Elongation at break | D882 | % | 400 |
| Elastic modulus | D882 | MPa (psi) | 413 (60,000) |
| Dielectric constant | D150 | $60-10^8$ Hz | 12.0 |
| Power/dissipation | D150 | – | 0.036 |
| Surface resistivity | D257 | – | Conductive |
| Volume resistivity | D257 | – | Conductive |
| Dielectric strength (DF1400 only) | D149 | V/mil | 4200 |
| Surface arc resistivity | D495 | | Does not arc |
| Continuous service temperature, °C (°F) | | −188 to 260 (−400 to 500) | |

*Note*: CHEMFILM DF1400 and DF1471 are multilayer cast PTFE films. The DF1400 has a nonconductive, fully fused, PTFE core coated with a conductive fluoropolymer formulation designed to be applied over jacketed wire and cable using standard tape-wrap equipment. The DF1471 is a fully conductive PTFE. *PTFE*, Polytetrafluoroethylene.
*Source*: Data from Saint Gobain, AFF-3020-1016 SGCS, 2016; <www.plastics-saint-gobain.com>.

**Table 20.13** **Textiles Coated International Cast Film, Properties.**

| Property | Data |
|---|---|
| Thickness range, mm (in.) | 0.0254–0.127 (0.001–0.005) |
| Yield, 1 mil thick film, $ft^2$/lb | 90 |
| Service temperature, °F (°C) | −328 to 600 (−200 to 315) |
| Tensile strength (typical[a]), psi (MPa) | 4800 (33.1) |
| Elongation (minimum), % | 400 |
| Dielectric constant ($60-10^8$ Hz) | 2.0 |
| Power/dissipation | $<0.0001$ |
| Surface resistivity | $1\times10^{18}$ |
| Volume resistivity | $>10^{15}$ |
| Dielectric strength, V/mil | 4200 |
| Surface arc resistance | Does not arc |

*Note*: The Cast Film is a pure multilayered film. Available grades include clear (unpigmented), pigmented (white, red, blue, yellow, and tan), surface treated for adhesive bonding, and thermally bondable on one side.
[a]0.002 in. (50 μm) film.
*Source*: Data from <www.textilescoated.com>, 2019.

**Table 20.14** **DW 203 Unsintered Polytetrafluoroethylene (PTFE) Film, Properties.**

| Property | ASTM Test Method | Data |
|---|---|---|
| Tensile strength, psi (MPa) | D882 | 1800 at 0.002 in. (12.4 at 0.05 mm) |
| Tensile strength, psi (MPa) | D882 | 1500 at 0.005 in. (12.4 at 0.127 mm) |
| Elongation, % | D6040 | 30 |
| Density, g/cm$^3$ | | 1.5 |
| Thickness range, in. (mm) | | 0.002–0.020 (0.0508–0.508) |

*Note*: DW 203 is an unsintered PTFE film. The polymer is fibrillated giving a good strength and low elongation in the machine direction.
*Source*: Data from <www.dewal.com>, 2017.

**Table 20.15** **FLUOROWRAP E125, Unsintered Polytetrafluoroethylene (PTFE) Film, Properties.**

| Property | Data | | | |
|---|---|---|---|---|
| Thickness, mil (mm) | 2.0 (0.051) | 3.0 (0.0076) | 4.0 (0.102) | 10.0 (0.254) |
| Density, g/cm$^3$ | 1.6 | | | |
| Tensile strength, MD[a], psi (MPa) | 1800 (17.2) | 1800 (17.2) | 1500 (15.2) | 800 (6.9) |
| Elongation at break, MD[a], % | 50 | 50 | 85 | 125 |

*Note*: E125 is a versatile, high-quality, unsintered PTFE film with excellent dielectric strength and conformability. These qualities make it suitable for a variety of electrical applications. Additional versions of this material with different properties and features, such as colors are available.
[a]Minimum.
*Source*: Data from <www.films.saint-gobain.com>, 2019.

**Table 20.16** **FluoroCal Calendered Unsintered Film, Properties.**

| Property | Value |
|---|---|
| Thickness range, in. (mm) | 0.002–0.005 (0.050–0.125) |
| Tensile strength, lb/in. | 30–75 |
| Elongation, % | 100–200 |
| Dielectric strength, V | 10,000–16,000 |
| Operating temperature, °F (°C) | − 100 to 500 (−73 to 260) |

*Note*: The films are produced by calendering and tensilizing PTFE skived films. They have high tensile strength and low elongation in the machine direction, low porosity, reduced permeability and reduced cold flow. *PTFE*, Polytetrafluoroethylene.
*Source*: Data from <www.actontech.com>, 2019.

formed, heat bonded, metalized, and laminated to many other materials, as well as used as a hot-melt adhesive. They exhibit the following important properties:

- chemical resistance, low permeability to liquid, gases, moisture, and organic vapors;
- electrical reliability, high-dielectric strength, very low dielectric constant and power factor, as well as no electrical tracking and charring;
- wide range of temperature service, high melting temperature, useful physical properties at cryogenic temperatures;
- mechanical toughness, high resistance to impact and tear;
- very high antistick properties and low frictional properties; and
- long-term weatherability.

Commercial products are available from several manufacturers and suppliers in a variety of grades with different properties and colors, as shown in Tables 20.17–20.19. Additional data and information are available from www.daikinchemicals.com, www.plastics.saint-gobain.com, and www.chemours.com.

## 20.3 Perfluoroalkoxy Polymer Films

There are essentially two types of perfluoroalkoxy polymers:

1. copolymer of perfluoropropylvinyl ether (PPVE) and TFE, which are known as PFA resin and
2. copolymer of perfluoromethylvinyl ether (PMVE) and TFE, which are known as MFA resin

In general, mechanical properties of PFA resin are very similar to those of PTFE resin within the temperature range from −200°C to +250°C (−328°F to +482°F). The mechanical properties of PFA and MFA resins at room temperature are practically identical; the difference becomes obvious only at elevated temperatures because of the lower melting point of MFA (280°C–290°C) as compared to that of PFA (300°C–310°C) [1]. For this reason, we are including the films made from them to this section.

**Table 20.17** CHEMFILM Fluorinated Ethylene Propylene (FEP) General-Purpose (GP) Fluoropolymer Film, Properties.

| Properties | Test Method | Units | Values |
|---|---|---|---|
| **General** | | | |
| Specific gravity | ASTM D792 | | 2.12–2.17 |
| Water absorption | | % | < 0.01 |
| **Mechanical** | | | |
| Tensile strength (RT) | ASTM D882 | MPa (psi) | 24 (3500) |
| Elongation at break (RT) | ASTM D882 | % | 300 |
| Tensile modulus (RT) | ASTM D882 | MPa (psi) | 480 (70,000) |
| Initial tear strength, 1 mil (25.4 μm) | ASTM D1004 | N ($lb_f$) | 2.2–2.7 (0.5–0.6) |
| Initial tear strength, 2 mil (50 μm) | ASTM D1004 | N ($lb_f$) | 4.9–5.3 (1.1–1.2) |
| Propagation strength, 1 mil (25.4 μm) | ASTM D1922 | N ($lb_f$) | 1.4–1.5 (0.33–0.32) |
| Propagation strength, 2 mil (50 μm) | ASTM D1922 | N ($lb_f$) | 2.4–2.7 (0.55–0.60) |
| Folding endurance (MIT) | ASTM D2176 | Cycles | 10,000 |
| **Electrical** | | | |
| Dielectric strength, 1 mil (50 μm) | ASTM D149 | kV/mm (V/mil) | 240 (6000) |
| Dielectric constant, 1 kHz | ASTM D150 | | 2.1 |
| Dissipation factor, 1 kHz | ASTM D150 | | 0.0003 |
| **Thermal** | | | |
| Melt point | | °C (°F) | 252–282 (485–540) |
| Continuous service temperature | | °C (°F) | 205 (400) |

| | | | |
|---|---|---|---|
| Specific heat | | J/(kg K) | 1172 |
| | | (BTU/(lb °F)) | (0.28) |
| Coefficient of thermal conductivity | ASTM D2863 | W/(m K) | 0.195 |
| | | BTU in./(h $ft^2$ °F) | (1.35) |
| Coefficient of linear thermal expansion | ASTM D696 | mm/(mm °C) | $9.9 \times 10^{-5}$ |
| | | in./(in. °F) | $(5.5 \times 10^{-5})$ |
| Flammability | UL 94 | | V-0 |
| Limiting oxygen index | ASTM D2863 | % | 95 |
| **Optical** | | | |
| Refractive index | ASTM D542 | | 1.341–1.347 |
| Solar transmission | ASTM E424 | % | 96 |

*Note*: The FEP GP film is a clear thermoplastic film that can be used for thermoforming, can be heat sealed, laminated, welded, metallized and used as a hot-melt adhesive. It is available as transparent and pigmented as well as cementable (surface treated) product.

*Source*: Data from AFF-1007R-116-SGCS, <www.plastics.saint-gobain.com>, 2016.

**Table 20.18 NEOFLON Fluorinated Ethylene Propylene Film, Characteristics.**

| Property | Value |
|---|---|
| Melting point, °C (°F) | 270 (518) |
| Maximum continuous service temperature, °C (°F) | 200 (392) |
| Heat resistance, rating | A |
| Chemical resistance, rating | A |
| Mechanical strength, rating | B |
| Weatherability, rating | A |
| Water-contact angle, degrees | 114 |
| Water vapor permeability[a], g/($m^2$ day) | 1.0 |
| Flame retardancy, UL 94 rating | 94V-0 |
| Transparency, rating | A |

Rating versus the general material "D": A > B > C > D.
[a]Film 0.1 mm (4 mil) thick.
*Source*: <www.daikinchemicals.com>.

## 20.3.1 Films From PFA Resins

PFA films are produced from the copolymer of TFE and PPVE (see above). These films offer the highest continuous use temperature 260°C (500°F) of any melt-processible fluoropolymer film, as well as many of the performance properties of PTFE in a clear, transparent form and can be heat sealed, thermoformed, welded, metalized, or laminated to a wide variety of materials. They offer a combination of excellent dielectric properties across a wide temperature and frequency range, the highest level of chemical and stress-crack resistance, excellent clarity, and weatherability. Other attractive features of PFA films are:

- exceptional dielectric strength,
- very low surface energy,
- very high antistick and release properties, and
- wide functional temperature range.

Commercial products are available from several manufacturers and suppliers in a variety of grades with different properties and colors, as shown in Tables 20.20–20.22. Additional data and information are available from www.daikinchemicals.com, www.plastics.saint-gobain.com, and www.chemours.com.

**Table 20.19 Teflon Fluorinated Ethylene Propylene (FEP) Fluoropolymer Film, Properties.**

| Property | Test Method | Units | Values |
|---|---|---|---|
| **General** | | | |
| Specific gravity | ASTM D1505 | $kg/m^3$ ($lb/ft^3$) | 2150 (134) |
| Coefficient of friction, kinetic, film to steel | ASTM D1894 | | 0.1–0.3 |
| **Mechanical** | | | |
| Tensile strength | ASTM D882 | MPa (psi) | 21 (3000) |
| Elongation at break | ASTM D882 | % | 300 |
| Yield point | ASTM D882 | MPa (psi) | 12 (1700) |
| Elastic modulus | ASTM D882 | MPa (psi) | 480 (70,000) |
| Impact strength | Chemours pneumatic impact tester | J/m<br>ft lb/in. | $7.7 \times 10^3$<br>144 |
| Tear strength—initial (Graves) | ASTM D1004 | N ($lb_f$) | 265 (270) |
| Tear strength—propagation (Elmendorf) | ASTM D1922 | N ($lb_f$) | 1.23 (125) |
| Folding endurance (MIT) | ASTM D2176 | Cycles | 10,000 |
| Bursting strength (Mullen) | ASTM D774 | kPa (psi) | 76 (11) |
| **Electrical** | | | |
| Dielectric strength, 1 mil (50 μm), 25°C short time | ASTM D149 Method A | kV/mm (V/mil) | 260 (6500) |
| Dielectric constant, 100 Hz to 1 MHz, 25°C | ASTM D150 | – | 2.0 |
| Dissipation factor, 1 kHz, 25°C | ASTM D150 | – | 0.0002–0.0007 |
| Volume resistivity −40°C to 240°C (−40°F to 464°F) | ASTM D257 | Ω cm | $>1 \times 10^{18}$ |

**Table 20.19** **Teflon Fluorinated Ethylene Propylene (FEP) Fluoropolymer Film, Properties.—cont'd**

| Property | Test Method | Units | Values |
|---|---|---|---|
| Surface resistivity −40°C to 240°C (−40°F to 464°F) | ASTM D257 | Ω/sq. | $1 \times 10^{16}$ |
| Surface arc resistance | ASTM D495 | s | >165 |
| **Chemical** | | | |
| Moisture absorption | – | % | <0.01 |
| Weatherability | Continuous exposure, Florida | – | No adverse effect after 20 years |
| Permeability, gas | ASTM D1434 | $cm^3/m^2 \cdot 24$ h atm | |
| Carbon dioxide | | | $25.9 \times 10^3$ |
| Hydrogen | | | $34.1 \times 10^3$ |
| Nitrogen | | | $5.0 \times 10^3$ |
| Oxygen | | | $11.6 \times 10^3$ |
| Permeability, vapor | ASTM E96 | $g/m^2$ day | |
| Acetone | | (g/100 $in.^2 \cdot 24$ h) | 14.7 (0.95) |
| Benzene | | | 9.9 (0.64) |
| Ethyl alcohol | | | 10.7 (0.69) |
| Hexane | | | 8.7 (0.56) |
| Water | | | 7.0 (0.40) |
| **Thermal** | | | |
| Melt range | ASTM D3418 | °C | 260–280 |
| | | °F | 500–536 |
| Continuous service temperature | – | °C | −240 to 205 |
| | | °F | −400 to +400 |

| Property | Test method | Unit | Value |
|---|---|---|---|
| Specific heat | – | J/kg K | 1172 |
| | | BTU/lb °F | 0.28 |
| Coefficient of thermal conductivity | Cenco-Fitch | W/(m K) | 0.195 |
| | | BTU in./h $ft^2$ °F | 135 |
| Dimensional stability | 30 min at 150°C (302°F) | | MD = 0.72% expansion<br>TD = −2.2% shrinkage |
| Flammability | ANS/UL 94 | | VTM-0 |
| LOI | ASTM D2863 | % | 95 |
| **Optical** | | | |
| Refractive index | ASTM D542 | | 1.341–1.347 |
| Solar transmission | ASTM E424 | % | 96 |
| **Other** | | | |
| Thickness | | mil | 0.50–20 |
| | | μm | 12.5–500 |
| Approx. area factor (depending on gauge) | | $ft^2$/lb | 36–180 |
| | | $m^2$/kg | 0.95–4.5 |

Types of Chemours Teflon FEP offered: Type A—general purpose; Type C—one-side cementable; Type C-20—both sides cementable. *LOI*, Limiting oxygen index.

*Source*: Data from C-10597 (9/17), The Chemours Company, <www.chemours.com>, 2019.

**Table 20.20 NEOFLON PFA Film, Characteristics.**

| Property | Value |
|---|---|
| Melting point, °C (°F) | 305 (581) |
| Maximum continuous service temperature, °C (°F) | 260 (500) |
| Heat resistance, rating | A |
| Chemical resistance, rating | A |
| Mechanical strength, rating | B |
| Weatherability, rating | A |
| Water-contact angle, degrees | 115 |
| Water vapor permeability[a], g/($m^2$ day) | 2.0 |
| Flame retardancy, UL 94 rating | 94V-0 |
| Transparency, rating (depending on grade) | A–C |

Rating versus the general material "D": A > B > C > D.
[a]Film 0.1 mm (4 mil) thick.
*Source*: Data from <www.daikinchemicals.com>.

## 20.3.2 Films From MFA Resins

MFA films are produced from the copolymer of TFE and PMVE (see above). As pointed out earlier, they have lower melting point than those from PFA resins but are still reported to offer the following properties [2]:

- outstanding clarity;
- performance from −254°C (−425°F) to 230°C (446°F);
- outstanding flex life and stress-crack resistance;
- outstanding antistick and release properties;
- excellent electrical properties; and
- excellent weatherability.

Examples for applications for MFA films are release films in aerospace industry, electrical/electronic industries, medical device industry, and chemical process industry.

Commercial products are available from several manufacturers and suppliers in a variety of grades with different properties and colors, as shown in Table 20.23.

**Table 20.21 CHEMFILM PFA Fluoropolymer Film.**

| Properties | Test Method | Units | Values |
|---|---|---|---|
| **General** | | | |
| Yield (1 mil) | | $m^2/kg$ ($ft^2/lb$) | 18 (90) |
| Specific gravity | ASTM D792 | | 2.12–2.18 |
| Water absorption, 24 h | | % | $<0.01$ |
| **Mechanical** | | | |
| Tensile strength (RT) | ASTM D882 | MPa (psi) | 21 (2250) |
| Elongation at break (RT) | ASTM D882 | % | 300 |
| Tensile modulus (RT) | ASTM D882 | MPa (psi) | 480 (70,000) |
| Initial tear strength, 2 mil (50 μm) | ASTM D1004 | N | 4.9–5.3 |
| | | $lb_f$ | (1.1–1.2) |
| Propagation tear strength, 2 mil (50 μm) | ASTM D1922 | N | 2.4–2.7 |
| | | $lb_f$ | (0.55–0.60) |
| Folding endurance (MIT) | ASTM D2176 | Cycles | 60,000 |
| **Electrical** | | | |
| Dielectric strength, 1 mil (50 μm) | ASTM D149 | kV/mm (V/mil) | 185 (4700) |
| Dielectric constant, 1 kHz | ASTM D150 | | 2.1 |
| Dissipation factor, 1 kHz | ASTM D150 | | 0.0005 |
| **Thermal** | | | |
| Melt point | | °C (°F) | 302–310 |
| | | | (575–590) |
| Continuous service temperature | | °C (°F) | 260 (500) |

**Table 20.21** CHEMFILM PFA Fluoropolymer Film.—cont'd

| Properties | Test Method | Units | Values |
|---|---|---|---|
| Specific heat | | J/(kg K) | 1172 |
| | | BTU/(lb °F) | (0.28) |
| Coefficient of thermal conductivity | ASTM D2863 | W/(m K) | 0.195 |
| | | BTU in./(h ft$^2$ °F) | (1.35) |
| Coefficient of linear thermal expansion | ASTM D696 | mm/(mm °C) | $9.9 \times 10^{-5}$ |
| | | in./(in. °F) | ($5.5 \times 10^{-5}$) |
| Flammability | UL 94 | | V-0 |
| Limiting oxygen index | ASTM D2863 | % | 95 |
| **Optical** | | | |
| Refractive index | ASTM D542 | | 1.35 |
| Solar transmission | ASTM E424 | % | 96 |

*Note*: The PFA film is a clear thermoplastic film that can be used for thermoforming, can be heat sealed, laminated, welded, metallized and used as a hot-melt adhesive. It is available as transparent and pigmented as well as cementable (surface treated) product. The following grades of PFA are offered: Type PG: General-purpose grade with good dielectric and optical properties, chemical and thermal resistance and weatherability. Type WF (mechanical welding grade) with the same properties as Type PG, but offering up to 25% cost savings.
*Source*: Data from AFF-XXXX-1116-SGCS, <www.plastics.saint-gobain.com>, 2016.

**Table 20.22** **Teflon PFA Fluoropolymer Film, Properties.**

| Properties | Test Method | Units | Values |
|---|---|---|---|
| **General** | | | |
| Density | ASTM D1505 | $kg/m^3$ ($lb/ft^3$) | 2150 (134) |
| Approx. area factor, 1 mil film | | $m^2/kg$ ($ft^2/lb$) | 18 (90) |
| Friction coefficient, kinetic, film to steel | ASTM D1894 | | 0.1–0.3 |
| **Mechanical** | | | |
| Tensile strength | ASTM D882 | MPa (psi) | 21 (3000) |
| Elongation at break | ASTM D882 | % | 300 |
| Yield point | ASTM D882 | MPa (psi) | 12 (1700) |
| Elastic modulus | ASTM D882 | MPa (psi) | 480 (70,000) |
| Impact resistance | Chemours pneumatic impact | J/m | $6.2 \times 10^4$ |
| | tester | in. lb/mil | 14 |
| Tear strength—initial (Graves) | ASTM D1004 | N (g) | 4.90 (500) |
| Tear strength—propagation (Elmendorf) | ASTM D1922 | N (g) | 0.74 (75) |
| Folding endurance (MIT | ASTM D2176 | Cycles | 100,000 |
| **Electrical** | | | |
| Dielectric strength, 1 mil (50 μm), 23°C short time | ASTM D149 Method A | kV/mm (V/mil) | 260 (6500) |
| Dielectric constant, 100 Hz to 1 MHz, 25°C | ASTM D150 | – | 2.0 |
| Dissipation factor, 100 Hz to 1 MHz, 25°C | ASTM D150 | – | 0.0002–0.0007 |
| Volume resistivity | ASTM D257 | Ω cm | $> 1 \times 10^{17}$ |
| −40°C to 240°C (−40°F to 464°F) | | | |
| **Chemical** | | | |
| Moisture absorption | – | % | < 0.02 |
| Permeability, gas | ASTM D1434 | $cm^3/m^2 \cdot 24$ h atm | |

**Table 20.22 Teflon PFA Fluoropolymer Film, Properties.—cont'd**

| Properties | Test Method | Units | Values |
|---|---|---|---|
| Carbon dioxide | | | $14 \times 10^3$ |
| Nitrogen | | | $2.0 \times 10^3$ |
| Oxygen | | | $6.7 \times 10^3$ |
| Permeability, vapor | ASTM E96 | $g/m^2$ day | 2 |
| Water | | $g/100\ in.^2 \cdot 24\ h$ | 0.13 |
| **Thermal** | | | |
| Melt range | ASTM D3418 | °C | 300–310 |
| | | °F | 575–590 |
| Continuous service temperature | – | °C | −240 to 205 |
| | | °F | −400 to +400 |
| Specific heat | – | J/kg K | 1172 |
| | | BTU/lb °F | 0.28 |
| Thermal conductivity | Cenco-Fitch | W/(m K) | 0.195 |
| | | $BTU\ in./h\ ft^2 \cdot °F$ | 1.35 |
| Dimensional stability | 30 min at 150°C (302°F) | MD = 1% shrinkage | |
| | | TD = 1% shrinkage | |
| LOI | ASTM D2863 | % | 95 |
| **Optical** | | | |
| Refractive index | ASTM D542 | | 1.350 |
| Solar transmission | ASTM E424 | % | 96 |

Types of Chemours Teflon PFA offered: Type LP—general purpose; Type CLP—one-side cementable; Type CLP-20—both sides cementable. *LOI*, Limiting oxygen index.

*Source*: Data from C-10134 (9/17), The Chemours Company, < www.chemours.com >, 2019.

**Table 20.23 Saint-Gobain Norton MFA Fluoropolymer Film, Properties.**

| Properties | Test Method | Units | Values |
|---|---|---|---|
| **General** | | | |
| Density | ASTM D1792 | $g/cm^3$ | 2.15 |
| Approx. area factor, 1 mil film | | $m^2/kg$ ($ft^2/lb$) | 18 (90) |
| Water absorption | | % | ≤0.030 |
| **Mechanical** | | | |
| Tensile strength, yield | ASTM D882 | MPa (psi) | 24–30 (3500–4375) |
| Elongation at break | ASTM D882 | % | 300 |
| Elastic modulus | ASTM D882 | MPa (psi) | 496 (72,000) |
| Folding endurance (MIT) | ASTM D2176 | Cycles | 6000 |
| **Electrical** | | | |
| Dielectric strength, 1 mil (50 μm) | ASTM D149 | kV/mm (V/mil) | 197 (5000) |
| Dielectric constant, 100 Hz, 25°C | ASTM D150 | – | 2.0 |
| Dissipation factor, 1 kHz | ASTM D150 | – | ≤0.0005 |
| Surface resistivity | ASTM D257 | Ω/sq. | $>1\times10^{15}$ |
| **Thermal** | | | |
| CTE, linear, at 20°C (68°F) | ASTM D696 | μm/m °C (μin/in. °F) | 99 (55) |
| Melting point | ASTM D3418 | °C (°F) | 280–290 (536–554) |
| Maximum service temperature, air | | °C (°F) | 230 (446) |
| Flammability, UL94 | | | V-0 |
| Oxygen index | ASTM D2863 | % | ≤95 |
| **Optical** | | | |
| Refractive index | ASTM D542 | | 1.35 |
| Haze | ASTM D1003 | % | 2.5–3.5 |
| Transmission, visible[a] | | % | 90 |

[a]Transparent, thickness not quantified.

*Source*: Data from Saint-Gobain Norton® MFA Fluoropolymer Film, <www.matweb.com>, 2019; C-10134 (9/17), The Chemours Company, <www.chemours.com>, 2019.

## 20.4 Ethylene Tetrafluoroethylene Copolymer Films

Films produced from copolymer of ethylene and TFE are transparent thermoplastic films that can be heat sealed, thermoformed, welded, metalized, or laminated to a wide variety of materials. They are chemically inert, mechanically tough, resistant to impact and tearing and exhibit excellent dielectric properties, and has very good nonstick and low frictional properties. In addition, they are inert to outdoor exposure and exhibit excellent long-term weatherability.

Commercial products are available from several manufacturers and suppliers in a variety of grades with different properties and colors, as shown in Tables 20.24–20.26.

Additional data and information are available from www.plastics.saint-gobain.com and www.chemours.com.

## 20.5 Polychlorotrifluoroethylene Films

Polychlorotrifluoroethylene (PCTFE) films are made from homopolymers or copolymers of chlorotrifluoroethylene and can be processed by

**Table 20.24 NEOFLON Ethylene Tetrafluoroethylene Film, Characteristics.**

| Property | Value |
|---|---|
| Melting point, °C (°F) | 260 (500) |
| Maximum continuous service temperature, °C (°F) | 150 (302) |
| Heat resistance, rating | B |
| Chemical resistance, rating | B |
| Mechanical strength, rating | A |
| Weatherability, rating | A |
| Water-contact angle, degrees | 96 |
| Water vapor permeability[a], g/(m$^2$ day) | 6.0 |
| Flame retardancy, UL 94 rating | 94V-0 |
| Transparency, rating | A |

Rating versus the general material "D": A > B > C > D.
[a]Film 0.1 mm (4 mil) thick.
*Source*: Data from <www.daikinchemicals.com>.

**Table 20.25** CHEMFILM Ethylene Tetrafluoroethylene Copolymer (ETFE)-E2 Fluoropolymer Film, Properties.

| Properties | Test Method | Units | Values |
|---|---|---|---|
| **General** | | | |
| Specific gravity | ASTM D792 | | 1.70–1.76 |
| Yield (1 mil) | | $m^2$/kg ($ft^2$/lb) | 22.5 (110) |
| Water absorption, 24 h | | % | < 0.03 |
| **Mechanical** | | | |
| Tensile strength (RT) | ASTM D882 | MPa (psi) | 48 (7000) |
| Elongation at break (RT) | ASTM D882 | % | 400 |
| Tensile modulus (RT) | ASTM D882 | MPa (psi) | 1100 (160,000) |
| Propagation strength, 1 mil (25 μm) | ASTM D1922 | N ($lb_f$) | > 2.9 (0.65) |
| **Electrical** | | | |
| Dielectric strength, 1 mil (50 μm) | ASTM D149 | kV/mm (V/mil) | 215 (5500) |
| Dielectric constant, 1 kHz | ASTM D150 | | 2.6 |
| Dissipation factor, 1 kHz | ASTM D150 | | < 0.0008 |
| **Thermal** | | | |
| Melt point | ASTM D3418 | °C (°F) | 250–270 (482–518) |
| Continuous service temperature | | °C (°F) | 165 (330) |
| Specific heat | | J/(kg K) | 2000 |
| | | BTU/(lb °F) | 0.46 |
| Coefficient of thermal conductivity | ASTM D2863 | W/(m K) | 0.238 |
| | | BTU in./(h $ft^2$ °F) | 1.65 |
| Flammability | UL 94 | | V-0 |
| Limiting oxygen index | ASTM D2863 | % | 31 |
| **Optical** | | | |
| Refractive index | ASTM D542 | | 1.4 |

*Note*: The ETFE (general-purpose) film is a clear thermoplastic film that can be used for thermoforming, can be heat sealed, laminated, welded, metallized. Available as transparent and pigmented (in blue, red, and custom colors) as well as cementable and bondable product with C-surface treatment on one or both sides.

*Source*: Data from AFF-3001-1116-SGCS, <www.plastics.saint-gobain.com>, 2016.

**Table 20.26** Tefzel Ethylene Tetrafluoroethylene Copolymer (ETFE) Fluoropolymer Film, Properties.

| Properties | Test Method | Units | Values |
|---|---|---|---|
| **General** | | | |
| Density | ASTM D1505 | $kg/m^3$ ($lb/ft^3$) | 1700 (106) |
| Approx. area factor, 1 mil | | $m^2/kg$ ($ft^2/lb$) | 20 (100) |
| Friction coefficient, kinetic, film to steel | ASTM D1894 | | 0.2–0.3 |
| **Mechanical** | | | |
| Tensile strength | ASTM D882 | MPa (psi) | 41 (6000) |
| Elongation at break | ASTM D882 | % | 300 |
| Yield point | ASTM D882 | MPa (psi) | 12 (1700) |
| Flex modulus | ASTM D882 | MPa (psi) | 830 (120,000) |
| Impact resistance | Chemours pneumatic impact tester | J/m<br>(in. lb/mil) | $6.2 \times 10^4$<br>14 |
| Tear strength—initial (Graves) | ASTM D1004 | N (g) | 4.90 (500) |
| Tear strength—propagation (Elmendorf) | ASTM D1922 | N (g) | 0.74 (75) |
| Folding endurance (MIT) | ASTM D2176 | Cycles | 50,000 |
| **Electrical** | | | |
| Dielectric strength, 1 mil (50 μm), 23°C short time | ASTM D149 Method A | kV/mm (V/mil) | 160 (4000) |
| Dielectric constant, 100 Hz to 1 MHz, 25°C | ASTM D150 | – | 2.6 |
| Dissipation factor, 100 Hz to 1 MHz, 25°C | ASTM D150 | – | 0.0007 |
| Volume resistivity, 170°C (338°F) | ASTM D257 | Ω cm | $>1 \times 10^{17}$ |

| **Chemical** | | | |
|---|---|---|---|
| Moisture absorption | – | % | < 0.02 |
| Permeability, gas | ASTM D1434 | $cm^3/m^2 \cdot 24$ h atm | |
| Carbon dioxide | | | $3.9 \times 10^3$ |
| Nitrogen | | | $0.5 \times 10^3$ |
| Oxygen | | | $1.6 \times 10^3$ |
| Permeability, vapor | ASTM E96 | $g/m^2$ day | 7.8 |
| Water | | $g/100\ in.^2 \cdot 24$ h | 0.5 |
| **Thermal** | | | |
| Melt point | ASTM D3418 | °C | 260–280 |
| | | °F | 500–536 |
| Continuous service temperature | – | °C | −100 to 150 |
| | | °F | −150 to +300 |
| Specific heat | – | J/kg K | 1172 |
| | | BTU/lb °F | 0.28 |
| Thermal conductivity | Cenco-Fitch | W/(m K) | 0.24 |
| | | BTU in./h $ft^2$ °F | 1.65 |
| Dimensional stability | 30 min at 150°C (302°F) | | MD = 1% shrinkage |
| | | | TD = 5% shrinkage |
| LOI | ASTM D2863 | % | 30 |
| **Optical** | | | |
| Refractive index | ASTM D542 | | 1.4 |
| Solar transmission | ASTM E424 | % | 90 |

*Note*: Tefzel ETFE films are transparent, thermoplastic films that can be heat sealed, thermoformed, heat bonded, welded, metallized, laminated (combined with dozens of other materials) and used as hot-melt adhesives. They are chemically inert, mechanically tough with a great resistance to impact and tearing, have excellent dielectric properties and possess long time weatherability. Types of Chemours Tefzel ETFE offered: Type LZ—general purpose; Type CLZ—one-side cementable; Type CLZ-20—both sides cementable. *LOI*, Limiting oxygen index.

*Source*: Data from C-10201 (9/17), The Chemours Company, <www.chemours.com>, 2019.

the usual manufacturing techniques, such as thermoforming, heat sealing, lamination, sheeting, and die cutting. They provide the highest moisture barrier of any clear thermoplastic film; are resistant to many chemicals, including strong oxidizing agents such as fuming oxidizing acids, liquid oxygen, and ozone and to sunlight; are nonflammable, plasticizer and stabilizer free; and can be processed within the same range of temperatures as most other thermoforming materials. The disadvantage of PCTFE films is that they are attacked by many organic materials [3].

Commercial products are available from several manufacturers and suppliers in a large variety of grades with different properties and features, as shown in Tables 20.27–20.36. Additional data and information are available from www.honeywell-aclar.com/product-information and www.daikinchemicals.com.

## 20.6 Ethylene-Chlorotrifluoroethylene Copolymer Films

Ethylene-chlorotrifluoroethylene (ECTFE) films are UV stable and optically clear with more than 90% transmission of light in the visible

**Table 20.27 NEOFLON Polychlorotrifluoroethylene Film, Characteristics.**

| Property | Value |
|---|---|
| Melting point, °C (°F) | 210 (410) |
| Maximum continuous service temperature, °C (°F) | 120 (248) |
| Heat resistance, rating | B |
| Chemical resistance, rating | B |
| Mechanical strength, rating | B |
| Weatherability, rating | A |
| Water-contact angle, degrees | 83 |
| Water vapor permeability[a], g/($m^2$ day) | 0.1 |
| Flame retardancy, UL 94 rating | 94V-0 |
| Transparency, rating | A |

Rating versus the general material "D": A > B > C > D.
[a]Film 0.1 mm (4 mil) thick.
*Source*: Data from <www.daikinchemicals.com>.

**Table 20.28 Honeywell Aclar Rx160, Mid-Range Barrier Film[a], Properties.**

| Properties[a] | Typical Value | | Test Method |
|---|---|---|---|
| | English | Metric | |
| Specific gravity | 2.11 | | ASTM D1505 |
| Yield | 21,896 in.$^2$/lb | 31.14 m$^2$/kg | |
| Haze | < 1% | | ASTM D1003 |
| Crystalline melting point | 412°F | 211°C | ASTM D4591 |
| Dimensional stability, 10 min at 300°F (149°C) | | | |
| MD | ≤ + 12% | | ASTM D1204 |
| TD | ≤ − 12% | | |
| **Tensile Strength** | | | |
| MD | 11,000–17,000 psi | 76–117 MPa | ASTM D882 |
| TD | 5500–7000 | 38–48 MPa | |
| **Elongation** | | | |
| MD | 80%–130% | | ASTM D882 |
| TD | 200%–275% | | |
| **Modulus, Secant** | | | |
| MD | 200,000–240,000 psi | 1379–1655 MPa | ASTM D882 |
| TD | 180,000–220,000 psi | 1241–1517 MPa | |
| Surface tension (treated side) | | ≥ 42 dyn/cm | ASTM D2578 |
| Water vapor–transmission rate | g/100 in.$^2$/day | g/m$^2$/day | ASTM F1249 |
| At 77°F (25°C)/60% RH | 0.0033 | 0.051 | |
| At 86°F (30°C)/60% RH | 0.0060 | 0.093 | |
| At 86°F (30°C)/65% RH | 0.0062 | 0.096 | |
| At 104°F (40°C)/60% RH | 0.0210 | 0.326 | |
| At 100°F (37.8°C)/100% RH | 0.0270 | 0.419 | |

[a]Film thickness 0.60 mil (15 μm), measured at 73°F (22.8°C).

**Table 20.29** Honeywell Aclar Rx20e, Mid-Range Barrier Film[a], Properties.

| Properties[a] | Typical Value | | Test Method |
|---|---|---|---|
| | English | Metric | |
| Specific gravity | 2.11 | | ASTM D1505 |
| Yield | 16,420 in.$^2$/lb | 23.35 m$^2$/kg | |
| Haze | < 1% | | ASTM D1003 |
| Crystalline melting point | 412°F | 211°C | ASTM D4591 |
| Dimensional stability, 10 min at 300°F (149°C) | | | |
| MD | ≤ + 12% | | ASTM D1204 |
| TD | ≤ − 12% | | |
| **Tensile Strength** | | | |
| MD | 11,000–16,000 psi | 76–110 MPa | ASTM D882 |
| TD | 6000–7200 | 41–50 | |
| **Elongation** | | | |
| MD | 90%–150% | | ASTM D882 |
| TD | 200%–275% | | |
| **Modulus, Secant** | | | |
| MD | 190,000–225,000 psi | 1310–1551 MPa | ASTM D882 |
| TD | 170,000–200,000 psi | 1241–1517 MPa | |
| Surface tension (treated side) | | ≥ 42 dyn/cm | ASTM D2578 |
| Water vapor–transmission rate | g/100 in.$^2$/day | g/m$^2$/day | ASTM F1249 |
| At 77°F (25°C)/60% RH | 0.0023 | 0.037 | |
| At 86°F (30°C)/60% RH | 0.0040 | 0.062 | |
| At 86°F (30°C)/65% RH | 0.0044 | 0.069 | |
| At 104°F (40°C)/75% RH | 0.0150 | 0.233 | |
| At 100°F (37.8°C)/100% RH | 0.0190 | 0.295 | |

[a]Film thickness 0.80 mil (20 μm), measured at 73°F (22.8°C).

**Table 20.30** Honeywell Aclar SupRx 900, Mid-Range Barrier Film[a], Properties.

| Properties[a] | Typical Value | | Test Method |
|---|---|---|---|
| | English | Metric | |
| Specific gravity | 2.11 | | ASTM D1505 |
| Yield | 14,595 in.$^2$/lb | 20.75 m$^2$/kg | |
| Haze | < 1% | | ASTM D1003 |
| Crystalline melting point | 412°F | 211°C | ASTM D4591 |
| Dimensional stability, 10 min at 300°F (149°C) | | | |
| MD | ≤ + 12% | | ASTM D1204 |
| TD | ≤ − 12% | | |
| **Tensile Strength** | | | |
| MD | 11,000–16,000 psi | 76–110 MPa | ASTM D882 |
| TD | 6000–7200 | 41–55 | |
| **Elongation** | | | |
| MD | 100%–150% | | ASTM D882 |
| TD | 200%–275% | | |
| **Modulus, Secant** | | | |
| MD | 190,000–225,000 psi | 1310–1551 MPa | ASTM D882 |
| TD | 165,000–210,000 psi | 1138–1448 MPa | |
| Surface tension (treated side) | | ≥ 42 dyn/cm | ASTM D2578 |
| Water vapor–transmission rate | g/100 in.$^2$/day | g/m$^2$/day | ASTM F1249 |
| At 77°F (25°C)/60% RH | 0.0018 | 0.028 | |
| At 86°F (30°C)/60% RH | 0.0030 | 0.047 | |
| At 86°F (30°C)/65% RH | 0.0038 | 0.059 | |
| At 104°F (40°C)/75% RH | 0.0130 | 0.202 | |
| At 100°F (37.8°C)/100% RH | 0.0170 | 0.264 | |

[a]Film thickness 0.90 mil (23 μm), measured at 73°F (22.8°C).

**Table 20.31 Honeywell Aclar UltRx 2000, High-Performance Barrier Film[a], Properties.**

| Properties[a] | Typical Value | | Test Method |
|---|---|---|---|
| | English | Metric | |
| Specific gravity | 2.11 | | ASTM D1505 |
| Yield | 6567 in.$^2$/lb | 9.34 m$^2$/kg | |
| Haze | < 1% | | ASTM D1003 |
| Crystalline melting point | 412°F | 211°C | ASTM D4591 |
| Dimensional stability, 10 min at 300°F (149°C) | | | |
| MD | ≤ + 6% | | ASTM D1204 |
| TD | ≤ − 6% | | |
| **Tensile Strength** | | | |
| MD | 7000–10,000 psi | 48–69 MPa | ASTM D882 |
| TD | 4500–7500 psi | 31–52 MPa | |
| **Elongation** | | | |
| MD | 150%–200% | | ASTM D882 |
| TD | 175%–250% | | |
| Modulus, secant | | | |
| MD | 170,000–200,000 psi | 1172–1379 MPa | ASTM D882 |
| TD | 170,000–200,000 psi | 1172–1379 MPa | |
| Surface tension (treated side) | | ≥ 42 dyn/cm | ASTM D2578 |
| Water vapor–transmission rate | g/100 in.$^2$/day | g/m$^2$/day | ASTM F1249 |
| At 77°F (25°C)/60% RH | 0.0012 | 0.019 | |
| At 86°F (30°C)/60% RH | 0.0025 | 0.039 | |
| At 86°F (30°C)/65% RH | 0.0019 | 0.029 | |
| At 104°F (40°C)/75% RH | 0.0066 | 0.102 | |
| At 100°F (37.8°C)/100% RH | 0.0077 | 0.119 | |

[a]Film thickness 2.00 mil (51 μm), measured at 73°F (22.8°C).

Table 20.32 **Honeywell Aclar UltRx 3000, High-Performance Barrier Film[a] Properties.**

| Properties[a] | Typical Value | | Test Method |
|---|---|---|---|
| | English | Metric | |
| Specific gravity | 2.11 | | ASTM D1505 |
| Yield | 4380 in.$^2$/lb | 6.23 m$^2$/kg | |
| Haze | < 1% | | ASTM D1003 |
| Crystalline melting point | 412°F | 211°C | ASTM D4591 |
| Dimensional stability, 10 min at 300°F (149°C) | | | |
| MD | ≤ + 6% | | ASTM D1204 |
| TD | ≤ − 6% | | |
| **Tensile Strength** | | | |
| MD | 6000–9000 psi | 41–62 MPa | ASTM D882 |
| TD | 4000–7000 psi | 28–48 MPa | |
| **Elongation** | | | |
| MD | 150%–225% | | ASTM D882 |
| TD | 175%–225% | | |
| **Modulus, Secant** | | | |
| MD | 160,000–200,000 psi | 1103–1379 MPa | ASTM D882 |
| TD | 160,000–200,000 psi | 1103–1379 MPa | |
| Surface tension (treated side) | | ≥ 42 dyn/cm | ASTM D2578 |
| Water vapor–transmission rate | g/100 in.$^2$/day | g/m$^2$/day | ASTM F1249 |
| At 77°F (25°C)/60% RH | 0.0008 | 0.012 | |
| At 86°F (30°C)/60% RH | 0.0016 | 0.025 | |
| At 86°F (30°C)/65% RH | 0.0015 | 0.023 | |
| At 104°F (40°C)/75% RH | 0.0040 | 0.062 | |
| At 100°F (37.8°C)/100% RH | 0.0057 | 0.088 | |

[a]Film thickness 3.00 mil (76 μm), measured at 73°F (22.8°C).

**Table 20.33 Honeywell Aclar UltRx 4000, Ultrahigh-Performance Barrier Film[a], Properties.**

| Properties[a] | Typical Value | | Test Method |
|---|---|---|---|
| | English | Metric | |
| Specific gravity | 2.11 | | ASTM D1505 |
| Yield | 3285 in.$^2$/lb | 4.67 m$^2$/kg | |
| Haze | < 1% | | ASTM D1003 |
| Crystalline melting point | 412°F | 211°C | ASTM D4591 |
| Dimensional stability, 10 min at 300°F (149°C) | | | |
| MD | ≤ + 5% | | ASTM D1204 |
| TD | ≤ − 25% | | |
| **Tensile Strength** | | | |
| MD | 6000–9000 psi | 41–62 MPa | ASTM D882 |
| TD | 4000–7000 psi | 28–48 MPa | |
| **Elongation** | | | |
| MD | 150%–225% | | ASTM D882 |
| TD | 150%–225% | | |
| **Modulus, Secant** | | | |
| MD | 160,000–200,000 psi | 1103–1379 MPa | ASTM D882 |
| TD | 160,000–200,000 psi | 1103–1379 MPa | |
| Surface tension (treated side) | | ≥ 42 dyn/cm | ASTM D2578 |
| Water vapor–transmission rate | g/100 in.$^2$/day | g/m$^2$/day | ASTM F1249 |
| At 77°F (25°C)/60% RH | 0.0006 | 0.009 | |
| At 86°F (30°C)/60% RH | 0.0010 | 0.016 | |
| At 86°F (30°C)/65% RH | 0.0009 | 0.015 | |
| At 104°F (40°C)/75% RH | 0.0031 | 0.048 | |
| At 100°F (37.8°C)/100% RH | 0.0042 | 0.065 | |

[a]Film thickness 4.00 mil (102 μm), measured at 73°F (22.8°C).

**Table 20.34** **Honeywell Aclar UltRx 6000, UltraHigh-Performance Barrier Film[a], Properties.**

| Properties[a] | Typical Value | | Test Method |
|---|---|---|---|
| | English | Metric | |
| Yield | 2180 in.$^2$/lb | 3.10 m$^2$/kg | |
| Specific gravity | 2.11 | | ASTM D1505 |
| Haze | < 1% | | ASTM D1003 |
| Crystalline melting point | 412°F | 211°C | ASTM D4591 |
| Dimensional stability, 10 min at 300°F (149°C) | | | |
| MD | ≤ + 2% | | ASTM D1204 |
| TD | ≤ − 2% | | |
| **Tensile Strength** | | | |
| MD | 4000–7000 psi | 28–48 MPa | ASTM D882 |
| TD | 4000–7000 psi | 28–48 MPa | |
| **Elongation** | | | |
| MD | 150%–225% | | ASTM D882 |
| TD | 150%–225% | | |
| **Modulus, Secant** | | | |
| MD | 160,000–200,000 psi | 1103–1379 MPa | ASTM D882 |
| TD | 160,000–200,000 psi | 1103–1379 MPa | |
| Surface tension (treated side) | | ≥ 42 dyn/cm | ASTM D2578 |
| Water vapor–transmission rate | g/100 in.$^2$/day | g/m$^2$/day | ASTM F1249 |
| At 77°F (25°C)/60% RH | 0.0003 | 0.0047 | |
| At 86°F (30°C)/65% RH | 0.0005 | 0.0078 | |
| At 104°F (40°C)/75% RH | 0.0020 | 0.0302 | |
| At 100°F (37.8°C)/100% RH | 0.0025 | 0.0380 | |
| Oxygen transmission rate | cm$^3$/100 in.$^2$/day atm | cm$^3$/m$^2$/day atm | ASTM D3985 |
| At 73°F (23°C)/65% RH | 0.63 | 9.77 | |
| At 104°F (40°C)/75% RH | 1.78 | 27.6 | |

[a]Film thickness 6.00 mil (152 μm), measured at 73°F (22.8°C).

**Table 20.35** Honeywell Aclar 33C, High-Performance Barrier Film[a], Properties.

| Properties[a] | Typical Value | | Test Method |
|---|---|---|---|
| | English | Metric | |
| Specific gravity | 2.12 | | ASTM D1505 |
| Yield | 1677 in.$^2$/lb | 2.38 m$^2$/kg | |
| Haze | <4.5% | | ASTM D1003 |
| Crystalline melting point | 403°F | 206°C | ASTM D4591 |
| Dimensional stability, 10 min at 300°F (149°C) | | | |
| MD | ≤ +2% | | ASTM D1204 |
| TD | ≤ −2% | | |
| **Tensile Strength** | | | |
| MD | 3000–4600 psi | 21–32 MPa | ASTM D882 |
| TD | 3000–4000 psi | 21–28 MPa | |
| **Elongation** | | | |
| MD | 50%–125% | | ASTM D882 |
| TD | 50%–125% | | |
| **Modulus, Secant** | | | |
| MD | 185,000–200,000 psi | 1276–1379 MPa | ASTM D882 |
| TD | 185,000–200,000 psi | 1276–1379 MPa | |
| **Tears Strength, Graves** | | | |
| MD | 425–525 g/mil | | ASTM D1004 |
| TD | 425–525 g/mil | | |
| Water vapor–transmission rate | 0.003 g/100 in.$^2$/day | 0.05 g/m$^2$/day | ASTM F1249 |
| At 100°F (37.8°C)/100% RH | | | |
| Thermal conductivity | | $4 \times 10^{-4}$ cal cm/cm$^2$ s °C | |
| Flammability | Nonflammable | | |
| Oxygen index | 100 | | ASTM D2863 |

[a]Film thickness 7.80 mil (198 μm), measured at 73°F (22.8°C).

**Table 20.36 NEOFLON Polychlorotrifluoroethylene DF-0015C1 Film, Characteristics.**

| Properties | Units | Values | Test Method |
|---|---|---|---|
| Average thickness | μm | 15 | ASTM D3595 |
| Haze | % | 0.30 | HAZE measure |
| Tensile strength | MPa | 47 | ASTM D3595 |
| Elongation | % | 130 | ASTM D3595 |
| Heat shrinkage rate (MD/TD) | % | + 8.1/ − 7.3 | ASTM D1249 150°C, 10 min |
| Water vapor permeability | $g/m^2/day$ | 0.48 | ASTM D1249 |
| Wettability[a] | μN | 500 | JIS K6768 |

*Note*: Additional products are available and are listed on www.daikinchemicals.com.
[a]Wettability can be improved by corona treatment.
*Source*: Data from tds-df-0015c1-E_ver01_Oct_2018, Daikin Industries, Ltd.

spectrum, while combining excellent fire-resistance properties with very good mechanical performance. They exhibit significantly lower permeation rate to water vapor than ethylene tetrafluoroethylene copolymer (ETFE) and provide the highest dielectric strength of all fluoropolymer films as well as an outstanding resistance to weathering and high energy radiation. Moreover, ECTFE films exhibit the highest abrasion resistance and the highest dielectric strength of all fluoropolymer films [4].

Commercial products are available from several manufacturers and suppliers in a large variety of grades with different properties and features, as shown in Tables 20.37–20.40. Additional data and information are available from www.plastics.saint-gobain.com, www.cshyde.com and www.westlakeplastics.com.

## 20.7 Polyvinylidene Fluoride Films

Polyvinylidene fluoride (PVDF) is strong and tough as reflected by its tensile properties and impact strength. Compared to many

**Table 20.37 Textiles Coated International Ethylene-Chlorotrifluoroethylene Copolymer Film, Properties.**

| Properties | Units | Test Method | Values |
|---|---|---|---|
| **General Properties** | | | |
| Specific gravity | | ASTM D792 | 1.68 |
| Yield (1 mil film) | $ft^2$/lb/mil | | 114 |
| | $m^2$/kg/25 μm | | 23.4 |
| Flammability | | UL 94 | V-0 |
| Water absorption | % | | < 0.01 |
| **Mechanical Properties** | | | |
| Tensile strength, 5 mil film | psi (MPa) | ASTM D882 | 6500 (45) |
| Elongation at break, 5 mil film | % | ASTM D882 | 250 |
| Tensile modulus | psi (MPa) | ASTM D882 | 200,000 (1380) |
| Initial tear strength | g/mil (N/mm) | ASTM D1004 | 500 (200) |
| Propagation tear strength | g/mil (N/mm) | ASTM D1922 | 1200 (480) |
| Folding endurance (MIT) | Cycles, ave. | ASTM D2176 | > 250,000 |
| **Thermal Properties** | | | |
| Continuous use temperature | °F (°C) | UL 746 B | 330 (165) |
| Melt point | °F (°C) | ASTM D3418 | 465 (240) |
| Coefficient of linear thermal expansion | in./(in. °F) | ASTM D696 | $9 \times 10^{-5}$ |
| **Electrical Properties** | | | |
| Dielectric strength, 1 mil (25 μm) film | V/mil (kV/mm) | ASTM D149 | 5500 (220) |
| Dielectric constant, 1 kHz | | ASTM D150 | 2.6 |
| **Optical Properties** | | | |
| Refractive index | | ASTM D542 | 1.44 |
| Solar transmission | % | ASTM E424 | 95 |
| **Product Offering** | | | |
| Width | in. (mm) | | Up to 60 (1524) |
| Thickness | mil (μm) | | 0.5–10 (12.5–250) |
| Standard colors | | | Clear |
| Surface treatments available | | | Chemical etching<br>Plasma treatment |

*Source*: Data from Ethylene-Chlorotrifluoroethylene Film for Use in High-Performance Applications, Data Sheet, TCI, <www.textilescoated.com>, 2019.

**Table 20.38 CHEMFILM Ethylene-Chlorotrifluoroethylene Fluoropolymer Film, Properties.**

| Property | Test Method | Value | Units |
|---|---|---|---|
| **General** | | | |
| Specific gravity | ASTM D792 | 1.68 | – |
| Yield | | 22 | $m^2/kg$ |
| | | 115 | $ft^2/lb$ |
| Water absorption, 24 h | | < 0.01 | % |
| **Mechanical** | | | |
| Tensile strength | ASTM D882 | 55 (8000) | MPa (psi) |
| Elongation at break | ASTM D882 | 250 | % |
| Tensile modulus | ASTM D882 | 1375 (200,000) | MPa (psi) |
| Initial tear strength | ASTM D1004 | 4.4 (450) | N (g/mil) |
| Propagating tear strength, 1 mil | ASTM D1922 | 11.6 (1200) | N (g/mil) |
| Folding endurance (MIT) | ASTM D2176 | > 250,000 | Cycles |
| **Electrical** | | | |
| Dielectric strength, 1 mil | ASTM D149 | 216 (5500) | kV/mm (V/mil) |
| Dielectric constant, 1 kHz | ASTM D150 | 2.55–2.63 | |
| Dissipation factor, 1 kHz | ASTM D150 | < 0.005 | |
| **Thermal** | | | |
| Melt point | ASTM D3418 | 240 (465) | °C (°F) |
| Continuous service temperature | | 150 (302) | °C (°F) |
| Flammability | UL 94 | V-0 | |
| Heat sealing temperature | °C (°F) | 215–260 (475–500) | |

*Source*: Data from AFF-1005R-1116-SGCS, <www.plastics.saint-gobain.com>, 2016.

**Table 20.39 Halar 500 LC Film, Properties.**

| Property | Value, Unit | Test Method |
|---|---|---|
| **Physical** | | |
| Specific gravity | 1.68 | ASTM D792 |
| Molding shrinkage-flow | 2.5% | ASTM D1238 |
| Water absorption (equilibrium) | < 0.10% | ASTM D570 |
| **Mechanical** | | |
| Tensile strength | 47.0 MPa | ASTM D638 |
| Elongation at break | 250% | ASTM D638 |
| Yield strength | 30.0 MPa | ASTM D638 |
| Tensile modulus | 1660 MPa | ASTM D638 |
| Coefficient of friction, versus itself dynamic | | ASTM D1894 |
| Static | 0.20 | |
| | 0.20 | |
| Taber abrasion resistance, 1000 cycles, 500 g, CS-17 Wheel | 5.00 mg | ASTM D4060 |
| **Impact** | | |
| Notched izod impact | | ASTM D256 |
| − 40°C, 3.20 mm | 210 J/m | |
| 23°C, 3.20 mm | No break | |
| **Hardness** | | |
| Rockwell hardness (*R*-scale) | 90 | ASTM D785 |
| Durometer hardness (shore *D*) | 75 | ASTM D2240 |
| **Thermal** | | |
| Melting temperature | 242°C | ASTM D3418 |
| Brittleness temperature | < 76.0°C | ASTM D746A |
| Thermal conductivity (40°C) | 0.15 W/m/K | ASTM C177 |
| Thermal stability, − 1% mass loss, $N_2$ | 405°C | TGA |
| Flammability | V-0 | UL 94 |

*Source*: Data from Product Information 320-F, Halar® 500LC Film, <www.cshyde.com>, 2019.

**Table 20.40** **Ethylene-Chlorotrifluoroethylene Film, Properties.**

| Property | Units | Test Standard | Result |
|---|---|---|---|
| **Mechanical** | | | |
| Tensile strength at yield | psi | ASTM D882 | 4940 |
| Tensile elongation at break | % | ASTM D882 | 220 |
| Tensile modulus | psi | ASTM D882 | 284,000 |
| Flexural modulus | psi | ASTM D790 | 240,000 |
| Tear strength, propagation | g/mil | ASTM D1004 | 513 |
| **Thermal** | | | |
| Continuous use temperature—UL | °F | | 302 |
| Heat deflection temperature at 66 psi | °F | ASTM D648 | 240 |
| Melt temperature, DSC | °F | – | 464 |
| Glass transition temperature | °F | ASTM D3418 | – |
| **Flammability** | | | |
| UL rating, UL 94 | – | | VTM-0 |
| Oxygen index (LOI) | % | ASTM D2863 | 60 |
| NBS smoke | Dmax | ASTM E662 | – |
| **Electrical** | | | |
| Surface resistivity | Ω | ASTM D257 | $>10^{16}$ |
| Dielectric strength at 0.003″ | V/mil | ASTM D150 | 2670 |
| Dielectric constant, 1 kHz | | ASTM D149 | 2.56 |
| Dissipation factor, 1 kHz | | ASTM D149 | 0.0025–0.0050 |
| **Other** | | | |
| Specific gravity | | ASTM D792 | 1.68 |
| Water absorption at 24 h | | ASTM D570 | <0.1 |
| Refractive index | – | – | – |
| Haze | % | ASTM D1003 | – |
| Area factor | in.$^2$/lb/mil | | 16,364 |

*LOI*, Limiting oxygen index.

*Source*: Data from Ethylene-Chlorotrifluoroethylene Film, Product Bulletin, <www.westlakeplastics>, 2019.

thermoplastics, PVDF has excellent resistance to creep and fatigue; yet in thin sections such as films, PVDF components are flexible and transparent. Additional advantages of PVDF films are

- resistance to many chemicals,
- resistance to UV and a good weathering, and
- piezoelectric and pyroelectric properties.

Commercial products are available from several manufacturers and suppliers in a large variety of grades with different properties and features, as shown in Tables 20.41–20.43. Additional data and information are available from www.textilescoated.com, integument.com and www.westlakeplastics.com.

## 20.8 THV Fluoroplastic Films

3M Dyneon THV fluoroplastic is a copolymer of TFE, HFP, and VDF. At this writing the manufacturer offers five solid resins with melting points 115°C (239°F), 130°C (266°F), 165°C (329°F), 185°C (365°F), and 225°C (437°F) that can be extruded into films. The films made from these polymers exhibit a very high clarity and a very good light transmission, excellent flexibility, chemical resistance, UV resistance, good weatherability, and dielectric properties. THV films are bondable to itself and to other substrates for multilayer constructions. An example of properties of the film, produced by Textiles Coated International (TCI), is in Table 20.44. Additional data and information are available from www.textilescoated.com and www.professionalplastics.com. It should be noted that only TCI produces films from the grade THV 500 and publishes their properties. No other record regarding commercial THV films other than the above were found.

## 20.9 Teflon AF Fluoroplastic Films

Teflon AF is a family of amorphous fluoroplastics similar to other amorphous polymers in optical clarity and mechanical properties, including strength. Like other amorphous thermoplastics, they also exhibit similar performance over a wide range of temperatures, excellent dielectric properties, and chemical resistance. The difference from most fluoroplastics is that they are soluble in selected solvents, have

**Table 20.41 Polyvinylidene Fluoride Film[a], Properties.**

| Properties | Units | Test Method | Values |
|---|---|---|---|
| **General Properties** | | | |
| Specific gravity | | ASTM D792 | 1.78 |
| Yield (1 mil film) | $ft^2/lb/mil$ | | 22.2 |
| | $m^2/kg/25\ \mu m$ | | 108 |
| Flammability | | UL 94 | V-0 |
| Water absorption | % | | $< 0.04$ |
| **Mechanical Properties** | | | |
| Tensile strength | psi (MPa) | ASTM D882 | 5000–7000 (35–48) |
| Elongation at break | % | ASTM D882 | 250 |
| Tensile modulus | psi (MPa) | ASTM D882 | 290,000 (2000) |
| Folding endurance (MIT) | Cycles, ave. | ASTM D2176 | $> 25,000$ |
| **Thermal Properties** | | | |
| Continuous use temperature | °F (°C) | UL 746 B | 300 (155) |
| Melt point | °F (°C) | ASTM D3418 | 330 (165) |
| Coefficient of linear thermal expansion | in./(in. °F) | ASTM D696 | $7 \times 10^{-5}$ |
| **Electrical Properties** | | | |
| Dielectric strength, 1 mil (25 μm) film | V/mil (kV/mm) | ASTM D149 | 4000 (160) |
| Dielectric constant, 1 kHz | | ASTM D150 | 7.5 |
| **Optical Properties** | | | |
| Refractive index | | ASTM D542 | 1.4 |
| Solar transmission | % | ASTM E424 | 90 |
| **Product Offering** | | | |
| Width | in. (mm) | | Up to 60 (1524) |
| Thickness | mil (μm) | | 1–10 (25–250) |
| Standard colors | | | Clear |
| Surface treatments available | | | Chemical etching<br>Plasma treatment |

[a]Based on Kynar KF.

*Source*: Data from Polyvinilidene Fluoride Film Used in High-Performance Applications; TCI, <www.textilescoated.com>, 2019.

**Table 20.42 Polyvinylidene Fluoride Film, Properties.**

| Property | Units | Test Standard | Result |
|---|---|---|---|
| **Mechanical** | | | |
| Tensile strength at yield | psi | ASTM D882 | 7550 |
| Tensile elongation at break | % | ASTM D882 | 160 |
| Tensile modulus | psi | ASTM D882 | 250,000 |
| Flexural modulus | psi | ASTM D790 | 260,000 |
| Tear strength, propagation | g/mil | ASTM D1004 | 735 |
| **Thermal** | | | |
| Continuous use temperature—UL | °F | | 265 |
| Heat deflection temperature at 66 psi | °F | ASTM D648 | 244 |
| Melt temperature, DSC | °F | - | 329−338 |
| Glass transition temperature | °F | ASTM D3418 | – |
| **Flammability** | | | |
| UL rating, UL 94 | – | | VTM-0 |
| Oxygen index (LOI) | % | ASTM D2863 | 43 |
| NBS smoke | Dmax | ASTM E662 | – |
| **Electrical** | | | |
| Surface resistivity | Ω | ASTM D257 | $>10^{16}$ |
| Dielectric strength at 0.003″ | V/mil | ASTM D150 | 1930 |
| Dielectric constant, 1 kHz | | ASTM D149 | 8.15−10.46 |
| Dissipation factor, 1 kHz | | ASTM D149 | 0.005−0.019 |
| **Other** | | | |
| Specific gravity | | ASTM D792 | 1.78 |
| Water absorption at 24 h | | ASTM D570 | 0.01 |
| Refractive index | – | – | 1.42 |
| Haze | % | ASTM D1003 | – |
| Area factor | in.$^2$/lb/mil | – | 15,480 |

*LOI*, Limiting oxygen index.
*Source*: Data from Polyvinylidene Fluoride Film, Product Bulletin, <www.westlakeplastics>, 2019.

**Table 20.43** **Fluorogrip PV Polyvinylidene Fluoride (PVDF) Kynar Film, Properties.**

| Property | Test Method | Metric Value | Metric Unit | English Value | English Unit |
|---|---|---|---|---|---|
| **General** | | | | | |
| Specific gravity | ASTM D792 | 1.7−1.79 | | 1.77−1.79 | |
| Yield (1 mil film) | | 22 | $m^2/kg$ | 107 | $ft^2/lb$ |
| Flammability | UL 94 | V-0 | | V-0 | |
| Water absorption (24 h) | ASTM D570 | 0.02 | % | 0.02 | % |
| **Mechanical** | | | | | |
| Tensile strength | ASTM D638 | 49−55 | MPa | 7105−7975 | psi |
| Elongation at break | ASTM D638 | 50−250 | % | 50−250 | % |
| Tensile modulus | ASTM D638 | 1600−2200 | MPa | 232,000−319,000 | psi |
| Initial tear strength, 3 mil | | | | 1620 | lb/in. |
| Propagating strength, 1 mil | ASTM D1004 | | | 735 | g/mil |
| **Electrical** | | | | | |
| Dielectric strength, 1 mil | ASTM D149 | 12.4 | kV/mm | 310 | V/mil |
| Dielectric constant, 1 kHz | ASTM D150 | 6.9 | | 6.9 | |
| Dissipation factor, 1 kHz | ASTM D150 | 0.013 | | 0.13 | |
| **Thermal** | | | | | |
| Melt point | ASTM D3418 | 165−168 | °C | 329−335 | °F |
| Continuous service A[a] | | 130 | °C | 265 | °F |
| Temperature S[a] | | 159 | °C | 315 | °F |
| Heat sealing temperature | | 220−235 | °C | 425−450 | °F |
| Degradation temperature | | | | | |
| A[a] | | 164 | °C | 325 | °F |
| S[a] | | 178 | °C | 500 | °F |

[a]*A*, Acrylic adhesive; *S*, silicone adhesive.

*Source*: Data from FluoroGrip® PV(PVDF) Kynar, Technical Data Sheet, <www.integument.com>, 2019.

**Table 20.44** THV 500[a] Fluoropolymer Extruded Films, Properties.

| Properties | Units | Test Method | Values |
|---|---|---|---|
| **General Properties** | | | |
| Specific gravity | | ASTM D792 | 1.98 |
| Area yield | $ft^2$/lb/mil | | 97 |
| | $m^2$/kg/25 μm | | 20.3 |
| Flammability | | UL 94 | V-0 |
| Water absorption | % | | $< 0.01$ |
| **Mechanical Properties** | | | |
| Tensile strength | psi (MPa) | ASTM D882 | 4500 (30) |
| Elongation at break | % | ASTM D882 | 600 |
| Tensile modulus | psi (MPa) | ASTM D882 | 30,000 (210) |
| Folding endurance (MIT) | Cycles, ave. | ASTM D2176 | $> 100,000$ |
| **Thermal Properties** | | | |
| Continuous use temperature | °F (°C) | UL 746 B | 250 (120) |
| Melt point | °F (°C) | ASTM D3418 | 330 (165) |
| Coefficient of linear thermal expansion | $10^{-5}$/°C | ASTM D696 | 11.4 |
| **Electrical Properties** | | | |
| Dielectric strength, 1 mil (25 μm) film | V/mil (kV/mm) | ASTM D149 | 1500 (60) |
| Dielectric constant, 1 kHz | | ASTM D150 | 4.8 |

| Property | Unit | Test Method | Value |
|---|---|---|---|
| **Optical Properties** | | | |
| Refractive index | | ASTM D542 | 1.36 |
| Solar transmission | % | ASTM D1003 | 95 |
| Haze, 4 mil (100 μm) film | % | ASTM D1003 | 1 |
| **Product Offering** | | | |
| Width | in. (mm) | | Up to 60 (1524) |
| Thickness | in. | | 0.005 and 0.010 |
| | μm | | 25 and 250 |
| Standard colors | | | Clear |
| Surface treatments available | | | Chemical etching |
| | | | Plasma treatment |

[a]Based on 3 M Dyneon THV 500 Fluoroplastic resin.

*Source*: Data from THV Film for Use in High Performance Applications; TCI, <www.textilescoated.com>, 2019.

high gas permeability, high compressibility, high creep resistance, low surface energy, and low thermal conductivity. They have the lowest dielectric constant of any known fluoroplastic and the lowest index of refraction of any known polymer [5]. At this writing, two commercial grades of Teflon AF resins, AF 1600 and AF 2400, are offered.

Films from Teflon AF resins can be readily produced by melt extrusion or from solutions in perfluorinated solvents by spin coating, spray, and dip coating [5]. Typical properties of Teflon AF resins suitable for films are listed in Table 20.45. At this point, there were no official material data from the films found, but it can be assumed that these would be similar to the data for resins published by the manufacturer. Typical properties of the two currently available commercial resins are given in Table 20.45.

## 20.10 Polyvinyl Fluoride Films

As pointed out earlier in Chapter 5, Films From Polyvinyl Fluoride, the only known commercial producer and supplier of polyvinyl fluoride (PVF) for the last 60 years has been the DuPont Company under the trademark Tedlar. DuPont produces a variety of films and powdered resins based on PVF [6]. PVF films are offered in both unoriented and oriented forms and differ considerably in their properties.

### 20.10.1 Oriented Polyvinyl Fluoride Films

*Oriented film*, Tedlar is a biaxially oriented PVF film with unique properties, which include excellent resistance to weathering, outstanding mechanical properties, and inertness toward a variety of chemicals, solvents, and staining agents [7]. At this writing, oriented films are available in more than 150 standard variations of opaque, pigmented, translucent, and transparent, and wide range of mechanical properties [8].

### 20.10.2 Unoriented Polyvinyl Fluoride Films

*Unoriented film*, Tedlar SP is designed to provide excellent conformability to substrates while maintaining the good durability, color stability, chemical resistance, and ease of cleaning expected of PVF film. Tedlar SP also offers custom color capability not previously available in PVF film. It possesses high elongation and moderate yield stress. This grade of PVF is coated on a carrier film, which is removed before

**Table 20.45** **Teflon AF Fluoropolymer Films, Properties [5].**

| Properties | Test Method | Units | Values | |
|---|---|---|---|---|
| | | | AF 1600 | AF 2400 |
| **General** | | | | |
| Specific gravity | ASTM D792 | | 1.78 | 1.67 |
| Water absorption | | % | < 0.01 | < 0.01 |
| **Mechanical** | | | | |
| Tensile strength at 23°C (73°F) | ASTM D638 | MPa | 26.9 ± 1.5 | 26.4 ± 1.9 |
| Tensile strength at 220°C (428°F) | ASTM D638 | MPa | 7.7 ± 6.1 | 4.2 ± 1.8 |
| Elongation at break at 23°C (73°F) | ASTM D638 | % | 17.1 ± 5.0 | 7.9 ± 2.3 |
| Elongation at break at 220°C (428°F) | ASTM D638 | % | 89.3 ± 131 | 8.4 ± 4.1 |
| Yield strength at 23°C (73°F) | ASTM D638 | MPa | 27.4 ± 1.0 | 26.4 ± 1.9 |
| Yield strength at 220°C (428°F) | ASTM D638 | MPa | – | 8.7 ± 4.0 |
| Tensile modulus | ASTM D638 | GPa | 1.6 | 1.5 |
| Flexural modulus at 23°C (73°F) | ASTM D790 | GPa | 1.8 ± 0.1 | 1.6 ± 0.1 |
| Flexural modulus at 220°C (428°F) | ASTM D790 | GPa | 1.0 ± 0.1 | 0.7 ± 0.1 |
| Hardness durometer at 23°C (73°F) | ASTM D1706 | Shore D | 77 | 75 |
| Hardness durometer at 220°C (428°F) | ASTM D1706 | Shore D | 70 | 65 |
| Glass transition temperature, $T_g$ | ASTM D3418 | °C (°F) | 160 (320) ± 5 | 240 (464) ± 10 |
| **Optical** | | | | |
| Refractive index | ASTM D542 | | 1.31 | 1.29 |
| Optical transmission | ASTM D1003 | % | > 95 | > 95 |

**Table 20.45** Teflon AF Fluoropolymer Films, Properties [5].—cont'd

| Properties | Test Method | Units | Values | |
|---|---|---|---|---|
| | | | AF 1600 | AF 2400 |
| **Chemical** | | | | |
| Contact angle with water | ASTM D570 | Degrees | 104 | 105 |
| Critical surface energy | | dyn/cm | 15.7 | 15.6 |
| Volume coefficient of thermal expansion | ASTM E831 | ppm/°C (°F) | 260 (500) | 301 (572) |
| Permeability coefficient, gases | | Barrer | | |
| Water | | | 1142 | 4026 |
| Oxygen | | | 340 | 990 |
| Nitrogen | | | 130 | 490 |
| Carbon dioxide | | | – | 2800 |

shipping. This protects the PVF film during surface treatment for adhesion, slitting, and other handling steps. After lamination, by using conventional means, Tedlar SP film can be easily embossed and will maintain high pattern definition [9]. It can also be subjected to high levels of forming without significant recovery stresses.

### 20.10.3 Classification of Polyvinyl Fluoride Films

Thus in general, commercial PVF films are available in clear or pigmented forms at various *degrees of orientation*, surface gloss, and adhesion treatment. The degree of orientation is expressed by a number in the range 1–5, where 1 is the most oriented and 5 the least oriented. The oriented grade is referred to as Type 3 or 5. A special Type 1 film has controlled shrinkage. Thus the types of available Tedlar films according to the degree of orientations are as follows:

Type 2: High strength, high flex variety
Type 4: High elongation, high tear modification
Type 5: Minimally oriented
Types 8 and 9: Unoriented Tedlar SP

Surface characteristics of the films are identified by letters: Letter "A" (one-side adherable) and "B" (two-side adherable). The surface with good antistick properties is identified by letter "S".

Tedlar films are generally available in thicknesses ranging from 1.0 to 2.0 mil (25–50 μm), clear or pigmented. Pigmented films are available in a variety of colors, including black, which can be also conductive, when special carbon black is used.

Given the large number of different products, the manufacturer developed a system of codes, designating the features of each film [10]. Below are examples with explanations:

**Example 1:** Product Code for Oriented Tedlar: **TABNMJFP**

Letter "T" in the code indicates the film is Tedlar (oriented film)

"AB" describes the film; if AB = TR, then the film is transparent, if AB = WH, it is white.

"NM" relates to the thickness of the film; it ranges from 05 to 40; 10 refers to 0.001 in. (25 μm), 15 to 0.0015 in. (37.5 μm).

"J" describes the surface treatment for adhesion, such as "A", one-side treated, "B", two-side treated, and "S", untreated.

**Table 20.46 Tedlar Polyvinyl Fluoride (PVF) Films, General Properties.**

| | Property | Typical Value | Test Method | Test Conditions |
|---|---|---|---|---|
| Physical | Bursting strength, Mullen | 29–65 psi | ASTM D774 | 22°C (73°F) |
| | Coefficient of friction (film/metal) | 0.18–0.21 | ASTM D1894 | 22°C (73°F) |
| | Density | 1.37–1.72 g/cm$^3$ | ASTM D1505 | 22°C (73°F) |
| | Impact strength, Spencer | 10–20 lb/mil | ASTM D3420 | 22°C (73°F) |
| | Moisture absorption | $<$0.5% for most types | Water immersion | 22°C (73°F) |
| | Water vapor transmission | 9–57 g/m$^2$ day | ASTM E96 | 39.5°C (103°F) 80% RH |
| | Refractive index | 1.40 $n_D$ | ASTM D542 Abbe refractometer | 30°C (86°F) |
| | Tear strength, propagated, Elmendorf | 15–60 g/mil | ASTM D1922 | 22°C (73°F) |
| | Tear strength initial, Graves | 260–500 g/mil | ASTM D1004 | 22°C (73°F) |
| | Tensile modulus | 300–380 $\times$ 10$^3$ psi | ASTM D882 | 22°C (73°F) |
| | Ultimate tensile strength | 8–16 $\times$ 10$^3$ psi | ASTM D882 | 22°C (73°F) |
| | Ultimate elongation | 90%–250% | ASTM D882 | 22°C (73°F) |
| | Ultimate yield | 6000–4900 | ASTM D882 | 22°C (73°F) |

| | | | | |
|---|---|---|---|---|
| Chemical | Chemical resistance | No visible effect | 1-year immersion in | |
| | | Strength and appearance not affected | Acids | 25°C (77°F) |
| | | | Bases | 25°C (77°F) |
| | | | Solvents | 25°C (77°F) |
| | | | 2 h immersion in | Boiling |
| | | | Acids | Boiling |
| | | | Bases | Boiling |
| | | | Solvents | – |
| | | | Soil burial—5 years | |
| | Gas permeability | | | |
| | Carbon dioxide | 11 $cm^3$/100 $in.^2$ (24 h) (atm) (mil) | ASTM D1434 | 24°C (75°F) |
| | Helium | 150 $cm^3$/100 $in.^2$ (24 h) (atm) (mil) | ASTM D1434 | 24°C (75°F) |
| | Hydrogen | 58.1 $cm^3$/100 $in.^2$ (24 h) (atm) (mil) | ASTM D1434 | 24°C (75°F) |
| | Nitrogen | 0.25 $cm^3$/100 $in.^2$ (24 h) (atm) (mil) | ASTM D1434 | 24°C (75°F) |
| | Oxygen | 3.2 $cm^3$/100 $in.^2$ (24 h) (atm) (mil) | ASTM D3985 | 24°C (75°F) |
| | Vapor permeability[a] | | | |
| | Acetic acid | 45 g/(100 $m^2$) (h) (mil) | ASTM E96 modified | 24°C (75°F) |
| | Acetone | 10,000 g/(100 $m^2$) (h) (mil) | ASTM E96 modified | 24°C (75°F) |
| | Benzene | 90 g/(100 $m^2$) (h) (mil) | ASTM E96 modified | 24°C (75°F) |
| | Carbon tetrachloride | 50 g/(100 $m^2$) (h) (mil) | ASTM E96 modified | 24°C (75°F) |
| | Ethyl acetate | 1000 g/(100 $m^2$) (h) (mil) | ASTM E96 modified | 24°C (75°F) |
| | Ethyl alcohol | 35 g/(100 $m^2$) (h) (mil) | ASTM E96 modified | 24°C (75°F) |
| | Hexane | 55 g/(100 $m^2$) (h) (mil) | ASTM E96 modified | 24°C (75°F) |
| | Weatherability | Excellent | Florida exposure | Facing South at 45° to horizontal |

**Table 20.46 Tedlar Polyvinyl Fluoride (PVF) Films, General Properties.—cont'd**

| | Property | Typical Value | | Test Method | Test Conditions |
|---|---|---|---|---|---|
| Thermal | Aging | 3000 h | | Circulating air oven | 150°C (302°F) |
| | Heat sealability | Some varieties—see Heat Sealability Technical Bulletin | | | |
| | Linear coefficient of expansion | $2.8 \times 10^{-5}$ in./in./°F | | | |
| | Shrinkage | | | | |
| | (Type 2) MD and TD | 4% at 130°C (266°F) | | Air oven, 30 min | |
| | (Type 3) TD only | 4% at 170°C (338°F) | | Air oven, 30 min | |
| | (Type 4) TD only | 2.5% at 170°C (338°F) | | Air oven, 30 min | |
| | Temperature range | | | | |
| | Continuous use | −72°C to 107°C (−98°F to 225°F) | | | |
| | Short cycles or release (1−2 h) | | | | |
| | Zero strength | 260°C–300°C (500°F–570°F) | | Hot bar | |
| Electrical | | TTR20SG4 | TWH20BS3 | | |
| | Corona endurance (h) | 2.5 | 6.2 | ASTM Suggested Test Method | 60 cPs, 1000 V/mil |
| | Dielectric constant | 8.5 | 11 | ASTM D150 | 1 Kc at 22°C (72°F) |
| | Dielectric strength (kV/mil) | 3.4 | 3.5 | ASTM D150 | 60 cPs kV/mil |
| | Dissipation factor (%) | 1.6 | 1.4 | ASTM D150 | 1000 cPs, 22°C (72°F) |
| | | 2.7 | 1.7 | ASTM D150 | 1000 cPs, 70°C (158°F) |
| | | 4.2 | 3.4 | ASTM D150 | 10 Kc, 22°C (72°F) |
| | | 2.1 | 1.6 | ASTM D150 | 10 Kc, 70°C (158°F) |
| | Volume resistivity (Ω cm) | $4 \times 10^{13}$ | $7 \times 10^{13}$ | ASTM D257 | 22°C (72°F) |
| | | $2 \times 10^{10}$ | $1.5 \times 10^{11}$ | ASTM D257 | 100°C (212°F) |

*Note*: DuPont Tedlar PVF film has excellent resistance to hydrolysis. Strength, yield stress, and elongation are not measurably affected after 60 h exposure in 85 psig steam, at temperature 163°C (325°F).

[a]At partial pressure or vapor at given temperature.

*Source*: Data from S. Ebnesajjad, Polyvinyl Fluoride—Technology and Applications of PVF, Elsevier, Oxford, UK, 2013, p. 153.

**Table 20.47 Typical Properties of Tedlar SP Polyvinyl Fluoride (PVF) Highly Conformable Film for Aircraft Interiors.**

| | | Typical Value | |
|---|---|---|---|
| **Property** | **Test Method** | **2-mil Pigmented** | **1-mil Transparent** |
| Unit weight, $g/m^2$ | ASTM D4321 | 74 | 36 |
| Tensile strength, MPa (kpsi) | ASTM D882 | 37 (5.5) | 41 (6) |
| Elongation, % | ASTM D882 | 225 | 200 |
| Shrinkage, % at 150°C (302°F) | ASTM D1204 | 2 | 2 |
| Gloss, 85, 60, 20 degrees | Gardner | 9.5, 12, 3 | 6, 16, 3 |
| Haze, internal | Gardner | N/A | 2 |
| Haze total | Gardner | N/A | 44 |
| Color, Delta E | – | 1.0 | N/A |

*Note*: This is an unoriented PVF film that possesses high elongation and moderate yield stress. It is available as multilayer film without resorting to adhesives or heat sealing. The Tedlar SP manufacturing process allows the production of different combinations, such as pigmented base layer covered with an integral clear top layer. The resulting film can be laminated, formed and embossed or otherwise converted as though it were a monolayer film. These films are available as one-side adherable (A) or two-side adherable (B).
*Source*: Data from Highly Conformable Film for Aircraft Interiors, H-56691-1 (10/95), DuPont, 2019.

"F" describes the surface gloss; "G", glossy, "M" medium gloss, "L", low gloss, "S" satin. Letter "E" indicates enhance film for aircraft.

Letters for colors are, for example, "WH" = shell white, "WB" antique white, "RB" = royal blue, etc.

**Example 2:** Product Code for Tedlar SP: **TABNMJFP**

Letter "T" = Tedlar

"AB" again describes the film; same as Example 1, that is, "TR" = transparent, "WH" is white

"NM" describes the Film thickness (gauge) same as Example 1

"H" describes the surface gloss; "H", high gloss, "M", medium gloss, "L", low gloss, "S", satin

"P" means "carrier" or "no carrier"; "P" can be 8 (with carrier film) or 9 (without carrier film).

**Table 20.48** **Tedlar Polyvinyl Fluoride (PVF) Release Films, Properties.**

| Property | Values | | | | |
|---|---|---|---|---|---|
| | TTR10SG3 | TMR10SM3 | TWH10SS3 | TWH20SS3 | TTR20SG4 |
| Nominal thickness, μm (mil) | 25 (1.0) | 25 (1.0) | 25 (1.0) | 51 (2.0) | 51 (2.0) |
| Gloss | High | Medium | Satin | Satin | High |
| Color | Clear | Translucent | White | White | Clear |
| Specular gloss 60 degrees | 65 | 8 | 15 | 22 | 65 |
| Surface roughness, min. avg, μm (μin.) | 0.18 (7) | 0.64 (25) | 0.25 (10) | 0.21 (8) | 0.25 (10) |
| Tenacity, min[a] | | | | | |
| MPa | 90 | 62 | 83 | 62 | 62 |
| kpsi | 13 | 9 | 12 | 9 | 9 |
| Web strength[a] | | | | | |
| N/m, width | 2275 | 1575 | 2100 | 3150 | 3150 |
| lb/in., width | 13 | 9 | 12 | 18 | 18 |
| Elongation (MD[a]), min, % | 95 | 100 | 75 | 110 | 125 |
| Approximate yield | | | | | |
| $m^2$/kg | 28.6 | 28.6 | 24.4 | 12.2 | 14.3 |
| $ft^2$/lb | 140 | 140 | 119 | 60 | 70 |

*Note*: These films are also available with one-side treated to accept adhesives or other bonding materials for the fabrication of specialty release laminates.

[a]Room temperature.

*Source*: Data from DuPont Tedlar® Polyvinyl Fluoride (PVF) Films —Composite Release Applications, K-26868 (05/13), DuPont, 2013.

**Table 20.49** Tedlar Films for Specialty Release Laminates, Physical and Thermal Properties.

| Description<br>Designation | Units | 1.0 mil UV Screening Transparent Type 3<br>TUT10BG3 | 1.0 mil Transparent Type 3<br>TTR10BG3 | 1.5 mil Low-Gloss White Type 3<br>TWH15BL3 | 2.0 mil Satin White Type3<br>TWH20BS3 | Test Method |
|---|---|---|---|---|---|---|
| **Physical Properties** | | | | | | |
| Area factor | $ft^2/lb$ ($m^2/kg$) | 140 (28.7) | 140 (28.7) | 87 (17.8) | 60 (12.3) | – |
| Ultimate tensile strength, min (MD) | kpsi (MPa) | 13 (90) | 13 (90) | 8 (55) | 9 (62) | ASTM D882 |
| Tensile modulus (MD) | kpsi (MPa) | 310 (2138) | 301 (2075) | 305 (2103) | 385 (2655) | ASTM D882 |
| Ultimate elongation, min (MD) | % | 95 | 95 | 90 | 110 | ASTM D882 |
| Bursting strength | psi/mil (MPa/m) | 56.9 (15,446) | 48.1 (13,057) | 28.9 (7845) | $>$34.7 ($>$9420) | ASTM D774, Mullen |
| Tear strength, propagating (MD) | g/mil (kN/m) | 17.1 (6.6) | 19.2 (7.4) | 23.1 (8.9) | 46.2 (17.8) | ASTM D1922, Elmendorf |
| Tear strength, propagating (TD) | g/mil (kN/m) | 19.0 (7.3) | 17.4 (6.7) | 18.6 (7.2) | 26.6 (10.3) | ASTM D11922, Elmendorf |
| Tear strength, initial (MD) | g/mil (kN/m) | 373 (144) | 423 (163) | 333 (129) | 506 (195) | ASTM D1004, Graves |
| Tear strength, initial (TD) | g/mil (kN/m) | 435 (168) | 478 (185) | 264 (102) | 377 (146) | ASTM D1004, Graves |
| Impact strength | in.lb/mil (kJ/m) | 20.3 (90.3) | 17.5 (77.9) | 9.6 (42.7) | 16.1 (71.6) | ASTM D3420, Spencer |

**Table 20.49 Tedlar Films for Specialty Release Laminates, Physical and Thermal Properties.—cont'd**

| Description<br>Designation | Units | 1.0 mil UV Screening Transparent Type 3<br>TUT10BG3 | 1.0 mil Transparent Type 3<br>TTR10BG3 | 1.5 mil Low-Gloss White Type 3<br>TWH15BL3 | 2.0 mil Satin White Type3<br>TWH20BS3 | Test Method |
|---|---|---|---|---|---|---|
| Specific gravity | – | 1.37 | 1.39 | 1.46 | 1.71 | ASTM D1505 |
| Coefficient of friction, film/metal | – | 0.21 | 0.21 | 0.18 | 0.18 | ASTM D1894 |
| Coefficient of abrasion | – | – | – | 385 | – | ASTM D658 |
| Moisture absorption | % | $<0.5$ | $<0.5$ | $<0.5$ | $<0.5$ | ASTM D570 |
| Moisture vapor transmission | $g/m^2$ day | 30.1 | 30.2 | 24.5 | 16.9 | ASTM E96 |
| **Thermal Properties** | | | | | | |
| Aging in air | Hours to embrittlement | 3000 | 3000 | 3000 | 3000 | Oven at 300°C |
| Heat sealability | – | – | Some varieties—see Heat Sealability Technical Bulletin | | | |
| Linear coefficient of expansion (MD) | m/m K | $7.8\times10$ | $8.8\times10$ | $6.7\times10$ | $9.7\times10$ | ASTM D696 |
| Linear coefficient of expansion (TD) | m/m K | $8.1\times10$ | $7.1\times10$ | $8.0\times10$ | $8.3\times10$ | ASTM D696 |
| Shrinkage, max (TD) | % at °C | 6 at 150 | 5 at 170 | 5 at 170 | 5 at 170 | ASTM D1204 |
| Specific heat | cal/g °C (kJ/kg K) | 0.42 (1.76) | 0.24 (1.01) | 0.26 (1.09) | 0.25 (1.05) | DuPont 990 Thermal Analyzer |

*MD*, Machine direction; *TD*, transverse direction; *UV*, ultraviolet.

*Source*: Data from S. Ebnesajjad, Polyvinyl Fluoride—Technology and Applications of PVF, Elsevier, Oxford, UK, 2013, p. 161.

**Note:** More detailed discussion regarding this is in Appendix 5 (Tedlar Film Designation Guide).

Clearly, there are many products available from the manufacturer (DuPont) and several selected distributors in a large variety of grades with different properties, and for many applications, as shown in Tables 20.46–20.49.

## References

[1] J.G. Drobny, Technology of Fluoropolymers, second ed., CRC Press, Boca Raton, FL, 2009, p. 41.
[2] Saint Gobain Norton® MFA Fluoropolymer film, <www.matweb.com>, 2009.
[3] J.G. Drobny, Technology of Fluoropolymers, second ed., CRC Press, Boca Raton, FL, 2009, p. 48.
[4] CHEMFILM® ECTFE Fluoropolymer Film, Document AFF-1005R-1116-SGCS, <www.plastics.saint-gobain.com>, 2016.
[5] L.W. McKeen, Film Properties of Plastics and Elastomers, fourth ed., Elsevier, Oxford, UK, 2017, p. 352.
[6] S. Ebnesajjad, Fluoroplastics, second ed., Melt Processible Fluoropolymers, vol. 2, Elsevier, Oxford, UK, 2015, p. 319.
[7] S. Ebnesajjad, Polyvinyl Fluoride-Technology and Applications of PVF, Elsevier, Oxford, UK, 2013, p. 151.
[8] L.W. McKeen, Film Properties of Plastics and Elastomers, fourth ed., Elsevier, Oxford, UK, 2017, p. 353.
[9] S. Ebnesajjad, Polyvinyl Fluoride-Technology and Applications of PVF, Elsevier, Oxford, UK, 2013, p. 185.
[10] S. Ebnesajjad, Polyvinyl Fluoride-Technology and Applications of PVF, Elsevier, Oxford, UK, 2013, p. 157.

# 21 Applications for Commercial Fluoropolymer Films

This chapter covers the applications for commercial fluoropolymer films in some details. First, applications for individual types of films are listed and then a summary of the applications of these films in major industries and activities is given in Table 21.1.

## 21.1 Applications for Polytetrafluoroethylene Films

As pointed out earlier, polytetrafluoroethylene (PTFE) films are produced from granular resins (skived film), aqueous dispersions (cast films), and fine powders (unsintered films). Each type of these films offers a high service temperature, outstanding chemical resistance, no-stick and release properties, and very good dielectric and barrier properties (in some cases) as outlined in Chapter 9, Polytetrafluoroethylene Films—Typical Properties and Applications, and Chapter 20, Commercial Grades of Fluoropolymer Films.

### 21.1.1 Applications for Skived Polytetrafluoroethylene Films

Skived PTFE films are available in the thickness ranging from 0.0005 to 0.125 in. (0.0127 to 3.175 mm) and exhibit superior flatness with the tightest tolerances in the industry and are available in cementable/bondable surfaces.

Because of their following exceptional properties, they are used in a large number of applications:

*Release*: Release films in the production of high-temperature flex circuit materials; release film in the molding of composite structures such as vacuum bagging; release film and composite sheets in food baking and grilling.

**Applications of Fluoropolymer Films.** DOI: https://doi.org/10.1016/B978-0-12-816128-9.00021-0

*Heat resistance*: Substrate (backing) for heat-resistant pressure-sensitive tapes, films, fabric-reinforced films, laminated films and composite sheets and belts.

*Chemical resistance*: Chemical-resistant hose inner core, septa, cap and storage liners, and diaphragms.

*Dielectric properties*: Bonding film in the production of microwave circuit boards.

*Low frictional properties*: Low friction (antisqueak specialty) in automotive industry.

*Barrier properties*: Gas and liquid tight barrier in compensators, expansion joints, and bellows.

*Dielectric properties*: Fusible wrap in electrical/electronic cable; wire and cable especially for high temperature and sensitive equipment.

*Electrically conductive* versions: These can be used for static-dissipative applications such as static-dissipative layer within high static paint hoses and static-dissipative laminates.

### 21.1.2 Applications for Cast Polytetrafluoroethylene Films

Cast PTFE films are available as thin monolayer films or as monolayer films with pure PTFE films and as multilayer construction with layers of PFA and/or fluorinated ethylene propylene (FEP) and fluoropolymer blends. The individual layers can be clear, colored, or conductive. The multilayer films can be configured with different polymers or polymer blends and are inherently void and pinhole free and exhibit superior dielectric performance, drape, and conformability. Multilayer films with outside layers of FEP and/or PFA/MFA are heat weldable and heat sealable. Cementable and pressure-sensitive adhesion versions are available. Cast films are supplied in the thickness ranging from 0.00025 to 0.005 in. (0.00635 to 0.127 mm).

Because of their following exceptional properties, cast films are used in a large number of applications:

*Release*: Mold release films in composite-molding operations; release sheets and laminates in food processing and ceramic processing.

*Dielectric properties*: Films for dielectric insulation and in the production of microwave circuit boards.

*Barrier*: Barrier films for septa applications; barrier films for protective clothing conductive films and laminates; and backing for pressure-sensitive tapes.

*Heat resistance*: Substrate (backing) for heat-resistant pressure-sensitive tapes.

*Electrically conductive* versions can be used for static-dissipative applications such as static-dissipative laminates.

*Hydrophobic nature*: This nature of PTFE films makes them a suitable material for the use for surgical gowns, surgical towels, protective gowns, and disinfection equipment package materials.

Applications for cast PTFE films are in Figs. 21.1 and 21.2.

**Figure 21.1** PTFE roofing elements at the Munich Airport Center. *PTFE*, Polytetrafluoroethylene. Courtesy Skyspan Europe GmbH.

**Figure 21.2** Protective clothing from fluoropolymer laminates. Courtesy Saint Gobain Corporation.

### 21.1.3 Applications for Unsintered Polytetrafluoroethylene Films

Unsintered PTFE films are used primarily for *electrical insulation*, particularly for wire and cable applications [cable insulation, harness wrap, and as dielectric medium in coaxial cables; in gaskets; expansion joints; valve seal, diaphragm application; coil winding; *zinc–air batteries*; *metallized microwave antennas*; flexible liner for transfer systems; substrate (backing) for pressure-sensitive tapes; pharmaceutical packaging/lamination; seal and gasket designs, where low creep is required].

Other applications include *battery/fuel-cell components*. The most common use of unsintered thin films is in *plumber's thread seal tape*.

## 21.2 Applications for Fluorinated Ethylene Propylene Films

FEP films offer excellent chemical resistance and release properties and can also be easily fabricated. Compared to PFA, FEP films are lower in cost and temperature and have a low dielectric constant and dissipation with a continuous use temperature of 400°F (205°C) and are suitable for use in cryogenic and high-temperature applications (see Chapter 10: FEP Films, Typical Properties and Applications, and Chapter 20: Commercial Grades of Fluoropolymer Films).

Environmental growth chambers and *solar collectors*, sample bags and containers, diaphragms, gaskets, protective linings, *flexible printed circuit* and *flat-cable* applications, and *electret condenser microphone* are typical applications for FEP films.

## 21.3 Applications for Perfluoroalkoxy Resin Films

This includes films produced from copolymers of perfluoropropylvinyl ether and tetrafluoroethylene (TFE), generally known as *PFA* fluoropolymers and copolymers of perfluoromethylvinyl ether, and TFE, known as *MFA* fluoropolymers. Their properties are essentially identical with the exception of their melting temperatures; PFA fluoropolymer has a melting temperature 20°C (36°F) higher than an MFA polymer (see Chapter 11: Perfluoroalkoxy Resin Films—Typical Properties).

### 21.3.1 Applications to Perfluoroalkoxy Resin Films

These films offer the highest continuous use temperature of 260°C (500°F) of any melt-processable fluoropolymer film type. They offer many of the performance properties of PTFE films in a clear, transparent form.

In general, a PFA film offers a combination of these following characteristics:

- excellent dielectric properties across a wide temperature and frequency range;
- highest level of chemical and stress-crack resistance; and
- excellent clarity and weatherability.

Typical applications for PFA films include its use in *circuit-board fabrication*, *flat-cable*, and *electrical-insulation* applications. The high-temperature resistance and nonwetting surface of PFA make it an ideal material for use as a *high-temperature release film or bagging film* for composite manufactures.

### 21.3.2 Applications for MFA Films

MFA films offer a combination of excellent dielectric properties across a wide temperature and frequency ranges; chemical and stress-crack resistance similar to PFA, a continuous service temperature of 230°C (446°F), and the highest clarity of any fluoropolymer film (see Chapter 11: Perfluoroalkoxy Resin Films—Typical Properties, and Chapter 20: Commercial Grades of Fluoropolymer Films). Other properties of MFA film are outstanding flex life, stress-crack resistance, favorable antistick properties, and very good weatherability. MFA film is a clear, transparent product, which can be heat sealed, thermoformed, welded, metallized, or laminated to a wide variety of materials.

Typical applications of MFA films are in chemical process industry (tank linings, roll covers); in solar collectors; ultraviolet (UV)-protective films; release films in composite manufacture; and in circuit boards and as electrical insulations.

It should be noted that the melting temperature of MFA resin is 20°C (36°F) lower than that of the PFA resin (see Chapter 11: Perfluoroalkoxy Resin Films—Typical Properties).

## 21.4 Applications for Films from Copolymer of Ethylene and Tetrafluoroethylene

Copolymer of ethylene and tetrafluoroethylene (ETFE) material offers the outstanding performance of fluoropolymer films made from it over a temperature range from cryogenic to 165°C (330°F). The resistance to chemicals and weathering, low flammability, and stress-crack resistance make these films excellent candidates for a wide variety of applications. For details, see Chapter 12, ETFE Films—Typical Properties and Applications, and Chapter 20, Commercial Grades of Fluoropolymer Films.

ETFE films offer very good mechanical properties and tear resistance. The high light transmission ( > 95%) of ETFE films, when combined with their very good mechanical and chemical properties, provides excellent performance in *architectural roofing designs* and *greenhouses* ETFE films are also a suitable choice in *solar collector applications* due to their outstanding physical properties and high coverage factor resulting from their low density. Furthermore, ETFE films offer a low surface energy, making them a good choice for *release applications* for process temperatures up to 350°F (177°C). The low density of this material, when compared to FEP and PFA films, provides greater coverage per pound for composite-molding release-film applications. The low surface energy of ETFE films can also be utilized in applications such as *wall coverings and antigraffiti protection* for high traffic areas. The use for the ETFE film in an architectural application is in Fig. 21.3.

**Figure 21.3** Example of an ETFE roof. *ETFE*, Copolymer of ethylene and tetrafluoroethylene. Courtesy DPA.

ETFE films offer processors excellent *dimensional stability* for secondary processing such as thermoforming and welding. Products utilizing *surface treatment* technologies improve bondability characteristics for metalizing processes and adhesive coating.

## 21.5 Applications for Polychlorotrifluoroethylene Films

Polychlorotrifluoroethylene (PCTFE) films come in a variety of homopolymer and copolymer configurations, which enable them to be heat sealed, thermoformed, laminated, sheeted, or die cut. They provide the following features (see Chapter 13: PCTFE Films—Typical Properties and Applications, and Chapter 20: Commercial Grades of Fluoropolymer Films):

- Superior permeability resistance
- Superior clarity
- Excellent formability
- Resistance to most chemicals and solvents
- Flame resistance

Because of the preceding properties, they are used in the following applications:

- Electroluminescent (EL) panel overlays
- Blister packaging
- Durable labeling
- Electronics' encapsulation and protection

Major application for PCTFE is in specialty films for packaging in cases where there are high moisture-barrier demands such as in *pharmaceutical blister packaging* and *health-care markets*.

In *EL lamps* (Fig. 21.4), PCTFE film is used to encapsulate phosphor coatings, which provide an area light when electrically excited. The film acts as a water-vapor barrier, protecting the moisture-sensitive phosphor chemicals. EL lamps are used in aircraft, military, aerospace, automotive, business equipment applications, and in buildings.

**Figure 21.4** Electroluminescent lamp with PCTFE film. *PCTFE*, Polychlorotrifluoroethylene. Courtesy Honeywell.

Another use for PCTFE films is for *packaging of corrosion-sensitive military and electronic components*. Because of excellent electrical-insulation properties, these films can be used to protect sensitive electronic components, which may be exposed to a humid or harsh environment. They can be thermoformed to conform to any shape and detail. PCTFE films are also used to *protect the moisture-sensitive liquid crystal display* panels of portable computers [1].

PCTFE films can be laminated to a variety of substrates such as PVC, polyethylene terephthalate glycol, amorphous polyethylene terephthalate, or polypropylene. Metallized films are used for electronic-dissipative and moisture-barrier bags for sensitive electronic components (Fig. 21.5), packaging of drugs (Fig. 21.6), and medical devices (Fig. 21.7).

Other applications for PCTFE are in pump parts, transparent sight glasses, flowmeters, tubes, and linings in the chemical industry and for laboratory ware [2].

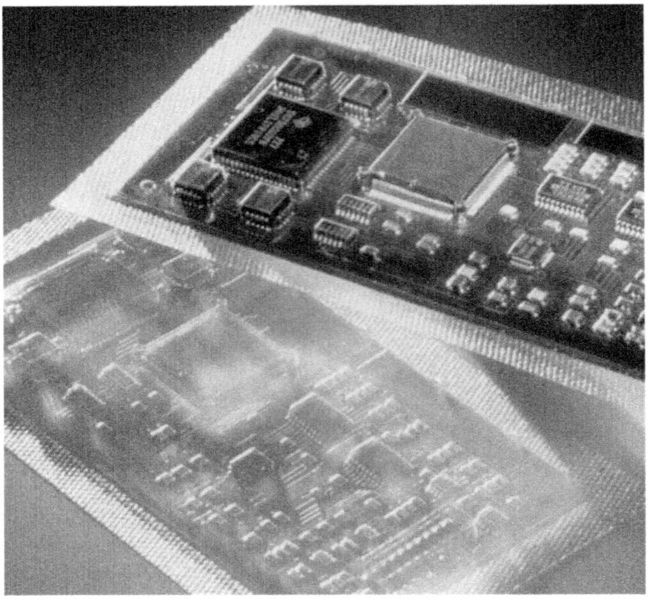

**Figure 21.5** Packaging of an electronic component with barrier film from PCTFE. *PCTFE*, Polychlorotrifluoroethylene. Courtesy Honeywell.

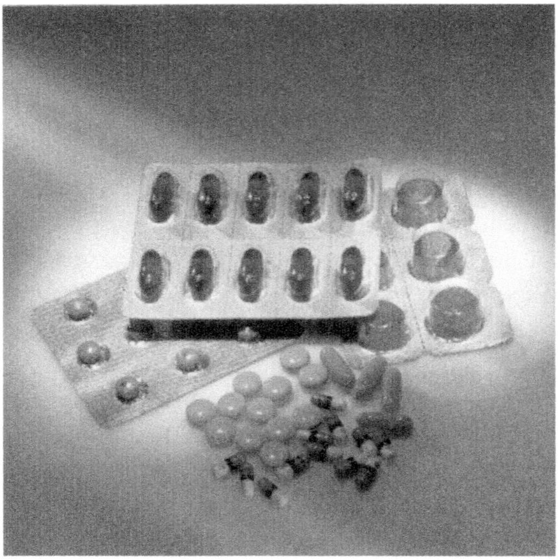

**Figure 21.6** Packaging of drugs with barrier film from PCTFE. *PCTFE*, Polychlorotrifluoroethylene. Courtesy Honeywell.

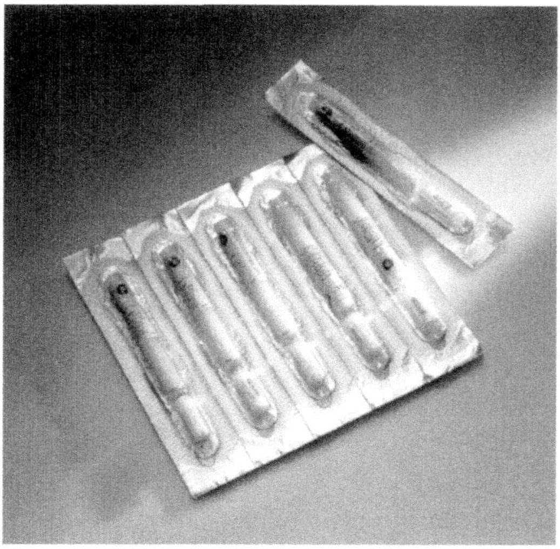

**Figure 21.7** Packaging of medical devices with barrier film from PCTFE. *PCTFE*, Polychlorotrifluoroethylene. Courtesy Honeywell.

## 21.6 Applications for Films from Copolymer of Ethylene and Chlorotrifluoroethylene

Copolymer of ethylene and chlorotrifluoroethylene (ECTFE) films offer excellent weatherability, chemical and abrasion resistance, and resistance to high-energy (ionizing) radiation and exhibit the highest dielectric strength of all fluoropolymer films (see Chapter 14: ECTFE Films—Typical Properties and Applications, and Chapter 20: Commercial Grades of Fluoropolymer Films). They are used in the *chemical-processing industry* as chemical tank lining, pump diaphragms, chlorine cells, in water treatment, spray-shielding application for pipe joints, and in semiconductor-processing environments. Because of their excellent weatherability, UV stability, abrasion resistance, and high light transmission, they are used in *outdoor protective applications*. Another major application of ECTFE films is in the back-sheet and front-sheet glazing of *photovoltaic (PV) panels* and as an external material for the back-sheet for protecting the PV module from the external environment for an extended period of time. ECTFE films are also used in *electrical and electronic industries* for tapes, cable insulation, printed circuits, capacitors, and flat-cable constructions. Other uses of these films are in *aircraft cabin interiors* and as *membranes in fuel cells.*

## 21.7 Applications for Polyvinylidene Fluoride Films

Polyvinylidene fluoride (PVDF) films have the following characteristics (details in Chapter 15: PVDF Films—Typical Properties and Applications, and Chapter 20: Commercial Grades of Fluoropolymer Films):

- Outstanding weatherability and resistance to UV radiation
- Resistance to many chemicals
- High fire resistance
- Excellent abrasion resistance
- High dielectric strength
- Continuous service up to 150°C (300°F)
- High antistick and low frictional properties
- Piezoelectricity and pyroelectricity

Due to their high resistance to most acids and solvents, PVDF films are used as a contact surface for the production, storage, and transfer of corrosive liquid. Applications include *chemical tank linings*, *pump diaphragms*, *water treatment*, and *chemical storage bags*.

Due to the outstanding dielectric performance, fire resistance, and high solar transmittance, PVDF films are well suited for use in the back-sheet and front-sheet glazing of *PV panels*. They are used extensively as an external material for the back-sheet, protecting the PV module from the external environment for an extended period of time. Other applications are *release films*, *piezoelectric films*, *pyroelectric films*, *fuel-cell seals*.

## 21.8 Applications for Films from Terpolymer of Tetrafluoroethylene, HFP and VDF

The key features of THV (terpolymer of tetrafluoroethylene, HFP, and VDF) films are (details in Chapter 16: THV Films—Typical Properties, and Chapter 20: Commercial Grades of Fluoropolymer Films):

- Excellent optical clarity and light transmission
- Bondability with many elastomers and plastics
- Outstanding flexibility in comparison with other fluoroplastics
- Excellent permeation resistance
- Good resistance to chemicals
- Low temperature processability

THV films provide a combination of performance advantages, such as low processing temperature, ability to bond to elastomers and hydrocarbon-based plastics, flexibility, and optical clarity.

Its typical applications include automotive, chemical-processing, multilayer tubes and hoses, semiconductors, solar energy, optical-fiber and architectural and protective coatings.

## 21.9 Applications for Teflon AF Films

As pointed out earlier in Chapter 17, Teflon AF Films—Typical Properties, and Chapter 20, Commercial Grades of Fluoropolymer Films, Teflon AF film can be readily produced from the resins by melt extrusion or by spin-coating or dip-coating techniques from their solutions in suitable solvents.

The high gas permeability makes Teflon AF an excellent separation medium for gases and liquids. In addition, membranes and film are used as high volume barriers or gas transport media. Examples include jet fuel and inkjet printer ink degassing

The films are essentially transparent to microwaves and can therefore function as a "window" *for high-frequency antennas*. The low dielectric constant and dissipation factor may be advantageous in the construction of electronic devices including special *circuit boards* and *hybrid devices*. Another application of these films is as *electronic dielectric for the next generation of high-speed computer circuits*, *interlayer dielectric*, and *deep UV-resistant films and components*.

## 21.10 Applications for Polyvinyl Fluoride Films

Commercial polyvinyl fluoride (PVF) films (Tedlar) are available in clear film and a wide variety of opaque pigmented film versions (for details, see Chapter 19: Polyvinyl Fluoride Films—Typical Properties and Applications, and Chapter 20: Commercial Grades of Fluoropolymer Films). The color uniformity and fade resistance of pigmented films allow them to maintain their original appearance for years. Transparent PVF films allow long-term light transmission while minimizing cracking, yellowing, or hazing of the laminate. They add strength to indoor and outdoor fabrics and prevent them from becoming brittle or mildewed. Surface-treated films are used for *pressure-sensitive adhesive tapes* and for *lamination* with a large variety of substrates including cardboard, paper, flexible PVC, polystyrene, rubber, polyurethane. These laminates are used for wall coverings, aircraft cabin interiors, pipe covering, duct liners, etc. For covering metal and rigid PVC, the film is first laminated to flat, continuous metal or vinyl sheets using special adhesives, and then the laminate is formed into the desired shapes. Such laminates are used for exterior sidings of industrial and residential buildings. Other applications are *highway sound barriers*, *automobile truck and trailer siding*, *vinyl awnings*, and *backlit signs*. On metal or plastic, PVF surfaces serve as a primer coat for painting or adhesive joints. PVF films are used as a *release sheet for bag molding of composites* from epoxide, polyester, and phenolic resins and in the manufacture of *circuit boards*.

Oriented commercial PVF films are available in more than 150 standard opaque, translucent, and transparent films, offering a wide array of attractive standard and custom colors and finishes. The product line is continuously growing.

With thicknesses ranging from 12.5 μm (0.5 mil) to 50 μm (2 mil) or higher, these films allow a high degree of design flexibility with respect to surface texture and formability. They can be manufactured with a range of special physical properties tailored to specific processing and performance requirements.

A summary of the most common applications for PVF films [3–6] is as follows:

*Aircraft*

PVF films give airlines maximum design flexibility in creating passenger areas that are attractive, easy to clean, and scuff resistant. PVF films comply with FAA25, as well as ABD-0031 regulations, are lightweight, have excellent conformability, and can be embossed and printed. PVF films are available in an array of colors, which resist fading. If necessary, the colors can be matched long after the initial installation. Transparent PVF film is an excellent top layer of laminates for printed patterns.

Typical application in the aircraft interiors are interior ceiling and sidewall decorative panels, window shades, stow bins, lavatories and galleys, ceiling panels, personal service units, bulkhead partitions, insulation barriers and moisture barriers, cargo bin liners, aircraft wire markers, and composite noise panels. The application for PVF laminates in the aircraft interior is shown in Fig. 21.8 and a sample of an aircraft interior laminate in Fig. 21.9.

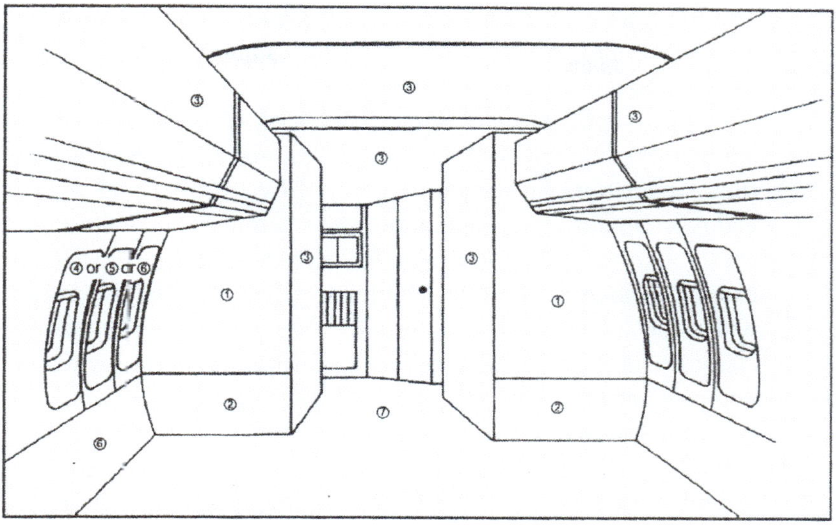


**Figure 21.8** Schematic diagram of an aircraft interior [3].
1—Panel construction with tapestry cover
2—Panel construction with wainscot cover
3—Panel construction with decorative plastic laminate
4—Formed thermoplastic or laminate
5—Formed aluminum with decorative plastic laminate
6—Composite laminate with wainscot cover
7—Panel construction with carpet cover

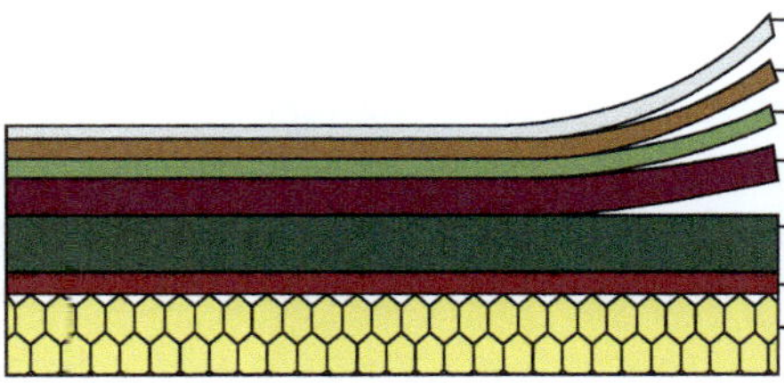


**Figure 21.9** Example of an aircraft interior laminate [3].

*Architectural*

PVF films are easily laminated to an array of architectural substrates, including metal, wood, cement, asphalt, vinyl, melamine composite, and fabrics. They provide superior protection that significantly prolongs their useful life and preserves their aesthetics for interior and exterior applications. They do not support bacterial growth and provide extra protection against contamination in hospitals, laboratories, clean rooms, and restaurants.

The inertness of PVF films resist staining, while their chemical stability lets the strongest cleaning agents be used without damaging the film or its substrate.

Typical applications using PVF films in architectural applications include wall coverings; ceiling and acoustical tiles; insulation jacketing; bagging film for thermal or acoustic materials; residential and commercial siding, trim, and accents; fiberglass-reinforced plastic (FRP) panels; formed or flat metal building panels; flexible laminates for air-inflated structures, canopies, awnings, and stadium domes; rigid composites fiberglass-reinforced polyester utility buildings and skylights; and conformable building panels.

*Forming*

Unlike gel coats, PVF films can be applied to FRP panels without ozone-depleting solvents. The application of films in place of paint eliminates much of the solvent commonly associated with high-performance paints.

*Graphics*

PVF films provide a tough, cleanable surface with excellent resistance to weather and harsh chemicals. They are the best protection against graffiti and grime. Transparent versions shield from UV radiation. Translucent and opaque pigmented films contain no plasticizers and therefore resist fading, chalking, and color shifts for years. PVF films can be used in transfer, reverse, or direct-printed applications with many water-based, solvent-based, or UV-curable inks. For heavy-duty graphics protection it is possible to print directly on white PVF film and cover it with a clear layer of PVF film.

Graphics applications using PVF films include

- gas pump skirts,
- flexible signs and awnings,
- billboards,
- highway signage, and
- labels.

*Solar applications*

In solar applications, PVF films are preferred as the backing sheet for PV modules due to their excellent strength, weather resistance, UV resistance, and moisture-barrier properties.

Solar applications using PVF films include

- PV modules and
- solar collectors.

Application of PVF in a solar panel is shown in Fig. 21.10.

**Figure 21.10** Example of a solar panel with PVF film. *PVF*, Polyvinyl fluoride. Courtesy DPA.

## 21.11 Applications for Fluoropolymer Films by Industries and Activities

The applications discussed in the previous sections of this chapter are summarized and listed by industries and activities in Table 21.1.

**Table 21.1 Applications of Fluoropolymer Films by Industry and/or Activity.**

| Industry/Activity | Film/Applications |
|---|---|
| Aerospace | ***Skived PTFE films:*** Release film in the molding of composite structures such as vacuum bagging<br>***Cast PTFE films:*** Mold release films in composite-molding operations for aircraft and space vehicles<br>***MFA films:*** Release films in manufacture of composite structures<br>***PCTFE films:*** EL lamps for aircrafts and aerospace equipment<br>***ECTFE films:*** Aircraft cabin interiors<br>***PVF films:*** Aircraft interiors (interior ceiling and sidewall decorative panels, window shades, stow bins, lavatories and galleys, ceiling panels, personal service units, bulkhead partitions, insulation barriers, moisture barriers, cargo bin liners, aircraft wire markers, composite noise panels, and release film for bag molding of composites) |
| Automotive | ***Skived PTFE films:*** Low friction (antisqueak specialty) in automotive industry and release film in the molding of composite structures for vehicles<br>***Cast PTFE films:*** Mold release films in composite-molding operations for vehicle parts<br>***MFA films:*** Release films in manufacture of composite structures<br>***PCTFE films:*** EL lamps for automotive applications<br>***THV films:*** Automotive interiors, layer in safety glass<br>***PVF films:*** Automobile truck and trailer siding |
| Chemical-processing industry | ***Skived PTFE films:*** Chemical-resistant hose inner core, septa, cap and storage liners and diaphragms; protective clothing<br>***Cast PTFE films:*** Barrier films for septa applications, storage liners and diaphragms, and protective clothing<br>***FEP films:*** Sample bags and containers, diaphragms, gaskets, and protective linings<br>***MFA films:*** Chemical-resistant tank linings and roll covers<br>***PCTFE films:*** Transparent sight glasses, flowmeters, tubes, and linings in the chemical industry; and laboratory ware and instruments<br>***ECTFE films:*** Chemical tank lining, pump diaphragms, chlorine cells, and in water treatment<br>***PVDF films:*** Chemical tank linings; pump diaphragms; water treatment; and chemical storage bags<br>***THV films:*** Multilayer tubes and hoses; gas sampling bags |

**Table 21.1** Applications of Fluoropolymer Films by Industry and/or Activity.—cont'd

| Industry/Activity | Film/Applications |
|---|---|
| Construction | ***ETFE films:*** Architectural roofing designs, greenhouses, and wall coverings<br>***PCTFE films:*** EL lamps in buildings<br>***ECTFE films:*** Outdoor protective applications<br>***THV films:*** Architectural and protective coatings<br>***PVF films:*** Wall coverings; ceiling and acoustical tiles; insulation jacketing; bagging film for thermal or acoustic materials; residential and commercial siding, trim, and accents; FRP panels; formed or flat metal building panels; flexible laminates for air-inflated structures, canopies, awnings, and stadium domes; rigid composites fiberglass-reinforced polyester utility buildings, and skylights; and conformable building panels |
| Consumer products | ***Cast PTFE films:*** Laminate sheets for home baking<br>***PVF films:*** Labels, flexible signs, laminates for silk-screened graphics |
| Defense | ***Skived PTFE films:*** Release films in the molding of composite structures for military vehicles<br>***Cast PTFE films:*** Release films in the molding of composite structures for military vehicles and so on; protective garments and tents<br>***PCTFE films:*** Films for packaging of corrosion-sensitive military components and EL lamps for the military<br>***PVF films:*** Release films in the molding of composite structures for military vehicles and so on; special laminates |

**Table 21.1 Applications of Fluoropolymer Films by Industry and/or Activity.—cont'd**

| Industry/Activity | Film/Applications |
|---|---|
| Electrical and electronic | ***Skived PTFE films:*** Bonding film in the production of microwave circuit boards; fusible wrap in electrical/ electronic cable; wire and cable, especially for high temperature and sensitive equipment; electrically conductive versions can be used for static-dissipative applications such as static-dissipative layer within high static paint hoses and static-dissipative laminates<br>***Cast PTFE films:*** Films for dielectric insulation; films in the production of microwave circuit boards; and electrically conductive versions can be used for static-dissipative applications such as static-dissipative laminates<br>***Unsintered PTFE films:*** *Electrical insulation*, particularly for wire and cable applications (cable insulation, harness wrap, and as dielectric medium in coaxial cables; metallized microwave antennas)<br>***FEP films:*** Flexible printed circuits; flat-cable applications; and electret condenser microphone<br>***PFA films:*** Circuit-board fabrication, flat cable, and electrical insulations<br>***MFA films:*** Circuit boards and electrical insulations<br>***PCTFE films:*** Protection the moisture-sensitive LCD panels of portable computers; EL lamps in several applications; metallized PCTFE films are used for electronic-dissipative and moisture-barrier bags for sensitive electronic components<br>***ECTFE films:*** Used in electrical and electronic industries for tapes, cable insulation, printed circuits, capacitors, and flat-cable constructions; in semiconductor-processing environments<br>***PVDF films:*** Piezoelectric and pyroelectric applications<br>***THV films:*** Semiconductor technology<br>***Teflon AF films:*** Special circuit boards and hybrid devices; "windows" for high-frequency antennas; electronic dielectric for the next generation of high-speed computer circuits; and interlayer dielectric<br>***PVF films:*** Circuit boards; piezoelectric and pyroelectric applications |

**Table 21.1** Applications of Fluoropolymer Films by Industry and/or Activity.—cont'd

| Industry/Activity | Film/Applications |
|---|---|
| Energy and environment | ***Unsintered PTFE films:*** Battery/fuel-cell components; zinc–air batteries<br>***MFA films:*** Solar collector applications<br>***ETFE films:*** Solar collector applications; antigraffiti protection for high traffic areas<br>***ECTFE films:*** *Membranes in fuel cells*; back-sheet and front-sheet glazing of photovoltaic panels and as an external material for the back-sheet to protect the photovoltaic module from the external environment for an extended period of time<br>***PVDF films:*** Back-sheet and front-sheet glazing of photovoltaic panels; external material for the back-sheet, protecting the photovoltaic module from the external environment for an extended period of time; and fuel-cell seals<br>***THV films:*** Solar energy applications<br>***Teflon AF films:*** Deep UV-resistant films and components<br>***PVF films:*** Backing sheet for photovoltaic modules; solar collectors; and protection against graffiti and grime |
| Engineering | ***Skived PTFE films:*** Gas and liquid tight barrier in compensators, expansion joints, and bellows<br>***Cast PTFE films:*** Laminate films for seals, compensators, and expansion joints<br>***Unsintered PTFE films:*** Flexible liner for transfer systems; seal and gasket designs where low creep is required; most common use is plumbers' thread seal tape<br>***PFA films:*** High-temperature release film or bagging film for composite manufacture<br>***MFA films:*** Release films in composite manufacture<br>***ETFE films:*** *Release applications* for process temperatures up to 350°F (177°C)<br>***PCTFE films:*** EL lamps in business equipment<br>***PVDF films:*** Release films<br>***PVF films:*** Backlit signs, gas pump skirts, shielding of UV radiation, billboards, highway signage, and highway sound barriers |

**Table 21.1 Applications of Fluoropolymer Films by Industry and/or Activity.—cont'd**

| Industry/Activity | Film/Applications |
|---|---|
| Food processing | ***Skived PTFE films:*** Release film and composite sheets for food baking and grilling<br>***Cast PTFE films:*** Release sheets and laminates in food processing, baking, and grilling |
| Medical | ***Cast PTFE films:*** Surgical gowns, surgical towels, protective gowns; disinfection equipment package materials |
| Packaging | ***Unsintered PTFE films:*** Pharmaceutical packaging/lamination<br>***PCTFE films:*** Pharmaceutical blister packaging; packaging of medical devices and packaging of corrosion-sensitive components |
| Safety and protection | ***Skived PTFE films:*** Protective clothing<br>***Cast PTFE films:*** Protective clothing<br>***MFA films:*** UV-protective films |
| Other | ***Skived PTFE films:*** Substrate (backing) for heat-resistant pressure-sensitive tapes, films, fabric-reinforced films, laminated films, and composite sheets and belts<br>***Cast PTFE films:*** Substrate (backing) for heat-resistant pressure-sensitive tapes<br>***Unsintered PTFE films:*** Substrate (backing) for pressure-sensitive tapes<br>***PVF films:*** Substrate (backing) for pressure-sensitive tapes; billboards; other graphics; and interiors of transit vehicles and passenger trains |

*ECTFE*, Copolymer of ethylene and chlorotrifluoroethylene; *EL*, electroluminescent; *ETFE*, copolymer of ethylene and tetrafluoroethylene; *FEP*, fluorinated ethylene propylene; *FRP*, fiberglass-reinforced plastic; *LCD*, liquid crystal display; *PCTFE*, polychlorotrifluoroethylene; *PTFE*, polytetrafluoroethylene; *PVDF*, polyvinylidene fluoride; *PVF*, polyvinyl fluoride; *THV*, terpolymer of tetrafluoroethylene; *HFP*, and *VDF*; *UV*, ultraviolet.

## References

[1] Aclar® Barrier Films, Honeywell International, Inc., <www.honeywell-aclar.com>, 2019.

[2] D.P. Carlson, W. Schmiegel, in: W. Gerhartz (Ed.), Ullmann's Encyclopedia of Industrial Chemistry, VCH Publishers, Weinheim, Germany, 1988, p. 412.

[3] S. Ebnesajjad, Polyvinyl Fluoride-Technology and Applications of PVF, Chapter 12, Elsevier, Oxford, UK, 2013.

[4] S. Ebnesajjad, Polyvinyl Fluoride-Technology and Applications of PVF, Chapter 13, Elsevier, Oxford, UK, 2013.

[5] S. Ebnesajjad, Polyvinyl Fluoride-Technology and Applications of PVF, Chapter 14, Elsevier, Oxford, UK, 2013.

[6] S. Ebnesajjad, Polyvinyl Fluoride-Technology and Applications of PVF, Chapter 15, Elsevier, Oxford, UK, 2013.

# Appendix 1: Major Fluoropolymer Film Manufacturers, Suppliers, and Distributors

| Company | Type | Film Product(s) | Web Page |
| --- | --- | --- | --- |
| Acton Technologies, Inc. | Manufacturer and supplier | Fluorocal calendered PTFE film | www.actontech.com |
| AGC Chemicals Americas, Inc. | Manufacturer and supplier | F-Clean ETFE film<br>Fluon ETFE | www.agcchem.com |
| American Durafilm Company | Distributor | Teflon PFA<br>Teflon FEP<br>DuPont Tedlar PVF | www.americandurafilm.com |
| The Chemours Company | Manufacturer and supplier | Teflon PFA<br>Teflon FEP<br>Tefzel ETFE | www.chemours.com |
| Daikin Industries | Manufacturer and supplier | Neoflon PFA<br>Neoflon FEP<br>Neoflon ETFE<br>Neoflon PCTFE | www.daikinchemicals.com |
| DeWal Industries, Inc. | Manufacturer and supplier | Skived PTFE films<br>Unsintered PTFE films<br>PTFE pressure sensitive tapes | www.dewal.com |
| DuPont | Manufacturer and supplier | DuPont Tedlar PVF<br>DuPont Tedlar S PVF | www.dupont.com |
| EMCO Industrial Plastics, Inc. | Distributor | DuPont Tedlar PVF<br>DuPont Tefzel ETFE<br>Arkema Kynar PVDF | www.emco.com |
| Honeywell International | Manufacturer and supplier | Aclar PCTFE | www.honeywell-aclar.com |

| | | | |
|---|---|---|---|
| Integument Technologies, Inc. | Manufacturer and supplier | Teflon FEP<br>Hyflon PTFE<br>Hyflon PFA<br>Hyflon MFA<br>Halar ECTFE<br>Kynar PVDF | integument.com |
| Saint-Gobain Performance Plastics-Coated Fabrics | Manufacturer and supplier | Chemfilm cast PTFE films<br>Laminates<br>Coated fabrics<br>PTFE tapes | www.chemfab.com |
| Saint-Gobain Tape Solutions | Manufacturer and supplier | CHR adhesive tapes | www.tapesolutions.saint-gobain.com |
| Saint-Gobain Performance Plastics-Specialty Films | Manufacturer and supplier | PTFE<br>FEP<br>PFA<br>ETFE<br>CTFE<br>ECTFE | www.plastics.saint-gobain.com |
| Textiles Coated International | Manufacturer and supplier | Cast PTFE<br>FEP<br>PFA<br>ETFE<br>PVDF<br>ECTFE<br>THV | www.textilescoated.com |
| Westlake Plastics | Manufacturer and supplier | Halar ECTFE<br>Kynar PVDF<br>Tefzel ETFE | www.westlakeplastics.com |

# Appendix 2: Trade Names of Common Commercial Fluoropolymers

| Trade Name | Company |
|---|---|
| 3M Dyneon Fluoroelastomers | 3M Dyneon |
| 3M Dyneon Fluoroplastic ETFE | 3M Dyneon |
| 3M Dyneon Fluoroplastic FEP | 3M Dyneon |
| 3M Dyneon Fluoroplastic PFA | 3M Dyneon |
| 3M Dyneon Fluoroplastic PTFE | 3M Dyneon |
| 3M Dyneon Fluoroplastic PVDF | 3M Dyneon |
| 3M Dyneon Fluoroplastic THV | 3M Dyneon |
| 3M Dyneon TFM | 3M Dyneon |
| Aclar | Honeywell |
| Aclar Films | Honeywell |
| Aclon | Honeywell |
| Aflas | AGC Chemicals Americas |
| Aflon | AGC Chemicals Americas |
| Algoflon PTFE | Solvay Specialty Polymers |
| Cefral Soft | Central Glass Co |
| Cytop | Asahi Glass Co. |
| DAI-EL | Daikin Industries |
| Daikin CTFE | Daikin America |
| Daikin EFEP | Daikin America |
| Daikin ETFE | Daikin America |
| Daikin PTFE | Daikin America |
| Fluon | AGC Chemicals Americas |
| Fluorobase T | Solvay Specialty Polymers |
| Fluoroplast | HaloPolymer Kirovo-Chepetsk |
| Fomblin PFPE | Solvay Specialty Polymers |
| Halar ECTFE | Solvay Specialty Polymers |
| Haleon | HaloPolymer Perm |
| Hylar | Solvay Specialty Polymers |
| Hyflon MFA | Solvay Specialty Polymers |

**cont'd**

| Trade Name | Company |
|---|---|
| Hyflon PFA | Solvay Specialty Polymers |
| Kalrez | DuPont |
| KF Polymer | Kureha Corporation |
| Kynar | Arkema |
| Kynar Flex | Arkema |
| Lumiflon | Asahi Glass Company |
| Nafion | Chemours |
| Neoflon ETFE | Daikin Industries |
| Neoflon FEP | Daikin Industries |
| Neoflon ETFE | Daikin Industries |
| Neoflon PFA | Daikin Industries |
| Polyflon PTFE | Daikin Industries |
| Polymist | Solvay Specialty Polymers |
| SKF | HaloPolymer, Kirovo-Chepetsk |
| Solef PVDF | Solvay Specialty Polymers |
| Tarflen | Zaklady Azotowe |
| Tedlar Films | DuPont USA |
| Teflon | Chemours |
| Teflon AF | Chemours |
| Teflon FEP | Chemours |
| Teflon NXT | Chemours |
| Teflon PFA | Chemours |
| Tefzel | Chemours |
| Viton | Chemours |
| Voltalef PCTFE | Arkema |
| Zeffle | Daikin America |
| Zonyl PTFE Micropowders | Chemours |

# Appendix 3: Acronyms and Abbreviations

| | |
|---|---|
| **ASTM** | American Society for Testing and Materials (now ASTM International) |
| **PFEVE** | poly(fluoroethylene vinyl ether) |
| **BMI** | bismaleimide |
| **CTE** | coefficient of thermal expansion |
| **CTFE** | chlorotrifluoroethylene |
| **DIN** | Deutsches Institut für Normung eV (German Institute for Standardization) |
| **DSC** | differential scanning calorimetry |
| **DTA** | differential thermal analysis |
| **E** | ethylene |
| **EB** | electron beam |
| **ECTFE** | copolymer of ethylene and chlorotrifluoroethylene |
| **ETFE** | copolymer of ethylene and tetrafluoroethylene |
| **FEP** | fluorinated ethylene-propylene (copolymer of tetrafluoroethylene and hexafluoropropylene) |
| **FEPM** | copolymer of tetrafluoroethylene and propylene |
| **FEVE** | fluorinated ethylene vinyl ether |
| **FKM** | fluorocarbon elastomer |
| **FMQ** | fluorosilicone |
| **FPM** | ISO designation for fluorocarbon elastomer of the FKM type (ASTM) |
| **FPU** | fluorinated polyurethane |
| **FTPE** | fluorinated thermoplastic elastomer |
| **FVE** | fluorovinyl ether |
| **FTIR** | Fourier-transform infrared spectroscopy |
| **Gy** | Gray (SI unit of radiation dose, larger, more widely used unit is kGy) |
| **HDPE** | high-density polyethylene |
| **HFIB** | hexafluoroisobutylene |
| **HFP** | hexafluoropropylene |

**HPFP** hydropentafluoropropylene
**IPN** interpenetrating network
**IR** infrared
**ISO** International Organization for Standardization
**JIS** Japanese Industrial Standard
**LAN** local area network
**LLDPE** linear low-density polyethylene
**LOI** limiting oxygen index
**MA** maleic anhydride
**MD** machine direction
**MDO** machine direction orientation
**MFA** copolymer of tetrafluoroethylene and tetrafluoroethylene and perfluoro(methyl vinyl ether)
**MFI** melt flow index
**MPa** megapascal (SI stress or pressure unit)
**Mrad** megarad (unit of dose, no longer used and superseded by SI unit Gy or kGy)
**MVE** methyl vinyl ether
**MWD** molecular weight distribution
**P** poise (unit of dynamic viscosity, superseded by SI unit of Pa s)
**Pa s** SI unit of dynamic viscosity (see poise)
$M_n$ number average molecular weight
$M_w$ weight average molecular weight
**NBS** National Bureau of Standards (in 1988 changed to NIST, see next)
**NIST** National Institute of Standards and Technology
**PA** polyamide
**PAVE** perfluoroalkyl vinyl ether
**PCTFE** polychlorotrifluoroethylene
**PDD** perfluoro-2,2-dimethyl dioxole
**PE** polyethylene
**PFA** perfluoroalkoxy polymer, copolymer of tetrafluoroethylene and perfluoro (propyl vinyl ether)
**PFOA** perfluorooctanoic acid
**PFOS** perfluorooctane sulfonate
**PI** polyimide
**PMVE** perfluoro(methyl vinyl ether)
**PP** polypropylene
**PPVE** perfluoro(propyl vinyl ether)

| | |
|---|---|
| **PTFE** | polytetrafluoroethylene |
| **PU, PUR** | polyurethane |
| **PVC** | polyvinyl chloride |
| **PVDF** | polyvinylidene fluoride |
| **PVF** | polyvinyl fluoride |
| **RF** | radiofrequency |
| **RoHS** | restriction of hazardous substances |
| **SBS** | styrene–butadiene–styrene block copolymer |
| **SI units** | International System of Units |
| **SKF** | fluorocarbon elastomer |
| **SSG** | standard specific gravity |
| $T_g$ | glass transition temperature |
| $T_m$ | crystalline melting point |
| **TD** | transverse direction (cross-machine direction) |
| **TDO** | transverse direction orientation |
| **TFE** | tetrafluoroethylene |
| **TGA** | thermogravimetric analysis |
| **THV** | terpolymer of tetrafluoroethylene, hexafluoropropylene, and vinylidene fluoride |
| **TPV** | thermoplastic vulcanizate |
| **UL** | Underwriters Laboratory |
| **UV** | ultraviolet |
| **VDF** | vinylidene fluoride |
| **VDI** | Verein Deutscher Ingenieure (The Association of Germen Engineers) |
| **VF** | vinyl fluoride |
| **VOC** | volatile organic compounds |

# Appendix 4: Glossary of Terms

## A

**Abrasion resistance**—Wear rate or abrasion rate measured by a number of methods, such as *Taber abrasion test* (ASTM D3389).

**Adhesive**—A material, usually polymeric, capable of forming permanent or temporary surface bonds with another material as is or after processing such as curing. The main classes of adhesives include hot melt, pressure-sensitive contact, UV cured and EB cured.

**Adherend**—A part to be covered by an adhesive and then joined into an adhesive joint.

**Adhesive bond strength**—The strength of a bond formed by joining two materials using an adhesive. Bond strength may be measured by peeling or shearing the two adherends (see above) using extensiometry.

**Amorphous polymer**—A polymer having a noncrystalline or amorphous supramolecular structure or morphology. Amorphous polymers may have some molecular order but usually are substantially less ordered than crystalline polymers and consequently have inferior mechanical properties.

**Annealing**—A process in which materials, such as plastic, glass, or metal, are heated and then cooled slowly. In plastics and metals, it is used to reduce stresses formed during fabrication.

**ASTM**—American Society for Testing and Materials is a nonprofit organization with the purpose of developing standards on characteristics and performance of materials, products, systems, and services and promoting the related knowledge. Now *ASTM International.*

**Average particle size**—The average diameter of solid particles as determined by various test methods.

## B

**Bar**—A metric (SI) unit of pressure, equal to $1.0 \times 10^6$ dyne/cm$^2$ or $1.0 \times 10^5$ Pa. It has the dimension of unit of force per unit of area and is used to denote the pressure of gases, vapors, and liquids.

**Biaxial orientation**—Orientation in which the material is drawn in two directions, usually perpendicular to one another. It can be either sequential or simultaneous. Commonly used for films and sheets.

**Blow molding**—The process of forming hollow articles by expanding a hot plastic element against the internal surfaces of a mold. In its simplest form the process comprises extruding a tube (parison) downward between opened halves of a mold, closing the mold, and injecting air to expand the tube, pinched on the bottom.

## C

**Calander**—A processing equipment used to form a thermoplastic material into a sheet or film. It consists of two or more steel (often heated) rolls with adjustable gap between them.

**Coalesce**—To combine particles into one body or to grow together.

**Coefficient of friction**—A number expressing the amount of frictional effect, usually expressed in two ways: static and dynamic.

**Cold flow (creep)**—Tendency of a material to flow slowly under load and/or over time.

**Comonomer**—A monomer reacting with a different monomer in a polymerization reaction, the result of which is a copolymer.

**Contact angle**—A measure of the ability of a liquid to wet solid surfaces. It expresses the relationship between the surface tension of a liquid and the surface energy of the surface on which the liquid rests. As the surface energy decreases, the contact angle increases.

**Corona treatment**—A method to render inert polymers more receptive to wetting by solvents, adhesives, coatings, and inks using high voltage discharge. The corona discharge oxidizes the surface, making it more polar.

**Cross-linking**—A reaction during which chemical links are formed between polymeric chains. The process can be carried out by chemical agents (e.g., organic peroxides), reactive sites on the polymeric chains, or by high-energy radiation (e.g., electron beam).

**Cryogenic**—Refers to very low temperatures, below about −150°C (−238°F).

**Crystalline melting point**—A temperature at which the crystalline portion of the polymer melts.

**Crystallinity**—A state of molecular structure attributed to the existence of solid crystals with a definite geometric form.

**Cure**—A process of changing the properties of a polymer by a chemical reaction (condensation, polymerization, or addition). In elastomers, it means mainly cross-linking or vulcanization.

## D

**Degradation**—Loss of or undesirable change in polymer properties as a result of aging, chemical reactions, wear, use, exposure, etc. The properties include color, size, and strength.

**Density**—The mass of any substance (gas, liquid, or solid) per unit volume at specified temperature and pressure.

**Dielectric constant**—The ratio of the capacitance assembly of two electrodes separated by a plastic insulating material to its capacitance when the electrodes are separated by air only.

**Dielectric heating**—The heating of polymeric materials by dielectric loss (see next) in a high-frequency electrostatic field.

**Dielectric loss**—A loss of energy evidenced by the rise in heat of a dielectric placed in an alternating electric field. It is usually observed as a frequency-dependent conductivity.

**Dielectric-loss factor**—The product of the dielectric constant and the tangent of the dielectric-loss angle for a material.

**Dielectric-loss tangent**—The difference between 90 degrees and the dielectric phase angle for a material.

**Dielectric strength**—Ability of a material to resist the passage of electric current. It is expressed in volts per thickness, which is required to break down through the thickness of a dielectric (insulation) material and create a puncture. ASTM D149 is the standard used to measure dielectric strength of plastic insulation materials.

***Differential scanning calorimetry* (DSC)**—The method to measure the heat flow to a sample as a function of temperature. It is used to measure specific heats, glass-transition temperatures, melting points, melting profiles, degree of crystallinity, degree of cure, purity, and more.

## E

**Elasticity**—The ability of a material to quickly recover its original dimensions after removal of the load that has caused the deformation.

**Elastomer**—A polymeric substance with elastic properties. Such a material can be stretched repeatedly at room temperature to at least twice its original length and upon immediate release of the stress will return with force to its approximate original length.

**ETFE**—Copolymer of ethylene and tetrafluoroethylene noted for an exceptional chemical resistance, toughness, and abrasion resistance.

**Electron beam cure**—A process using electron beam radiation to promote reactions in a polymeric materials, leading to cross-linking, polymerization, modification, and degradation.

**Electron beam (EB) radiation**—Ionizing radiation propagated by electrons, accelerated by very high voltage (typically kilovolts to megavolts). This radiation is used frequently for the processing of polymeric materials (see *electron beam cure*).

## F

**FEP**—Fluorinated ethylene propylene having excellent nonstick and nonwetting properties.

**Film formation**—A process in which a film is formed after solvent or water evaporates or due to a chemical reaction.

**Friction, dynamic**—Resistance to continued motion between two surfaces, also known as *sliding friction.*

**Friction, static**—Resistance to initial motion between two surfaces.

**Fuel cell**—An electrochemical energy-conversion device. It produces electricity from various external quantities of fuel (on the anode side) and oxidant (on the cathode side). These react in the presence of an electrolyte. Fuel cells are different from batteries in which they consume reactant, which must be replenished, while batteries store electrical energy chemically in a closed system.

**Fusion**—A process in which a continuous film or a solid body is formed by melting and flowing (coalesce) of polymer particles.

## G

***Glass-transition temperature*** **($T_g$)**—A point below which an amorphous polymer behaves as glass does—it is very strong and rigid, but brittle. Above this temperature, it exhibits leathery or rubbery behavior.

## H

**Heat buildup**—Heat generated within a polymeric material due to its viscoelasticity (hysteresis) and friction. It occurs during processing (mainly friction and kneading) and in service (mainly repeated cycling).

**Hysteresis**—Incomplete recovery of strain during the unloading cycle due to energy consumption. This energy is converted from mechanical to frictional energy (heat).

**HFP**—Hexafluoropropylene A monomer used for the production of FEP and other copolymers, such as THV, and of fluorinated elastomers.

## I

**Ionizing radiation**—Any electromagnetic or particulate radiation, which in its passage through matter is capable of producing ions directly or indirectly. Examples are electron beam and gamma radiation.

**Ionomer resins**—Modified polymers obtained by heating and pressing certain polymers containing carboxylic groups in the presence of metallic ions.

## L

**Laminate**—A product made by bonding together one or more layers of material or materials. It is frequently assembled by simultaneous application of heat and pressure. A laminate may consist of coated fabrics, metals, and films or it may be different combinations of these.

**Latex**—A stable dispersion of a polymeric substance (most frequently of an elastomer) in an essentially aqueous medium.

**Limiting oxygen index (LOI)**—LOI is defined as the required minimum percentage of oxygen in a mixture with nitrogen, which allows a flame to be sustained by an organic material such as a polymer.

## M

**Melt-processible polymer**—A polymer that melts when heated to its melting point and forms a molten material with definite viscosity value at or somewhat above its melting temperature. Such a melt can be pumped and should flow when subjected to shear rate using commercial-processing equipment, such as extruders and molding machines.

**MFA**—A copolymer of TFE and perfluoro(methyl vinyl ether) with properties similar to PFA; it has approximately 20°C lower melting temperature than PFA.

**Micron (micrometer)**—A unit of length equal to $1 \times 10^{-6}$ m. The *micrometer* (international spelling as used by the International Bureau of Weights and Measures; SI symbol: μm) or *micrometer* (American spelling), also commonly known by the previous name *micron*, is an SI derived unit of length.

**Modified PTFE**—Copolymer of TFE and of a small amount (less than 1%) of other perfluorinated monomers (e.g., perfluoroalkoxy monomer), exhibiting considerably improved physical properties, moldability, and much lower microporosity.

**Monomer**—A relatively simple compound, usually containing carbon and of a low molecular weight, which can react to form a polymer by combining with itself or with other similar molecules or compounds.

## N

**Nanometer**—A unit of length equal to $1 \times 10^{-9}$ m. Often used to denote the wavelength of radiation, especially UV and visible spectral region, and size of very small particles. Its symbol is nm.

**Newtonian fluid**—A term to describe an ideal fluid in which shear stress is proportional to shear rate with *viscosity* being the proportionality coefficient. In this case, viscosity is independent of shear rate in contrast to nonideal fluids, where viscosity is a function of shear rate. The latter represents non-Newtonian fluids that include paints and polymer melts.

## O

**Orientation**—A process of drawing or stretching of thermoplastic films and fibers in order to polymerize in the direction of stretching. While fibers are drawn in one direction, films may be drawn in one direction (uniaxially, either longitudinally or transversely) or two directions (biaxially). Oriented films and fibers have enhanced properties in the direction of stretching. The film will shrink in the direction of stretching when reheated without tension.

**Ozone**—Molecule consisting of three atoms of oxygen, that is, $O_3$.

## P

**Perfluorinated resin**—A polymer consisting of monomers where all main chain carbons are combined with fluorine atoms only (PTFE, FEP, PFA, and MFA).

**Permeability**—The capacity of a material to allow another substance to pass through it or the quantity of a specific gas or other substance, which passes through under specific conditions.

**PFA**—Copolymer of TFE with perfluoro(propyl vinyl ether), an engineering thermoplastic characterized by excellent thermal stability, release properties, low friction, and toughness. Its performance is comparable to PTFE with the difference that it is melt processible.

**Piezoelectricity**—The ability of some materials (notably crystals and certain ceramics) to generate an electric potential in response to applied mechanical stress. This may take the form of a separation of electric charge across the crystal lattice. If the material is not short-circuited, the applied charge induces a voltage across the material. The effect finds useful applications, such as the production and detection of sound, generation of high voltages, electronic frequency generation, microbalances, and ultrafine focusing of optical assemblies.

**PMVE**—Perfluoro(methyl vinyl ether), a monomer used for the production of MFA.

**Polymer fume fever**—An illness characterized by temporary flu-like symptoms caused by inhaling the products released during the decomposition of fluoropolymers, mainly PTFE. Tobacco smoke enhances the severity of this condition.

**Postcure**—A second cure at high temperatures enhancing some properties and/or removing decomposition products of the primary reaction.

**PPVE**—Perfluoro(propyl vinyl ether), a monomer used for the production of PFA.

**Prorad**—Radiation promoter, a compound promoting or enhancing the cross-linking reaction by high-energy (ionizing) radiation.

**Pyroelectricity**—The ability of certain materials to generate an electrical potential when they are heated or cooled. As a result of this change in temperature, positive and negative charges move to opposite ends through migration (i.e., the material becomes polarized) and hence, an electrical potential is established.

## R

**Radiation dose**—Amount of ionizing radiation energy absorbed by a material during irradiation. The unit of radiation dose is a gray (Gy), defined as 1 J/kg. In practical application a large unit, namely, kGy ($10^3$ Gy) is used. Previously used unit, no longer official since 1986, was megarad (Mrad), equal to 10 kGy. See *ionizing radiation.*

**Rheology**—A science that studies and characterizes the flow of polymers, resins, gums, and other materials.

## S

**Sintering**—A process in which particles are heated, softened, and coalesced, thus forming a continuous film or a solid body, used typically in the processing of PTFE.

**Specific gravity**—The ratio of the density of a substance to the density of a standard, usually water, for a liquid or solid, and air for a gas.

**Substrate**—Any surface to be coated by a coating or bonded by an adhesive.

**Surface energy**—The energy associated with the intermolecular forces at the interface between two media. The *surface energy* per unit area equals the surface tension, also called *free surface energy.* Solid materials can be divided into two categories: high or low surface energy. High-energy solids include metals and inorganic compounds, with typical values of 200–500 mN/m. Low-energy materials are generally organic materials, including polymers with values below 100 mN/m (see also *surface tension*).

**Surface resistance**—The surface resistance between two electrodes in contact with a material is the ratio of the voltage applied to the electrodes to that portion of the current between them, which flows through the surface layers.

**Surface tension**—The cohesive force at a liquid surface measured as a force per unit length along the surface or the work, which has to be done to extend the area of surface by a unit area, for example, by a square centimeter. It is an important factor in the wetting of solids by liquids and in the formation of adhesive bonds.

**Surfactant**—A widely used contraction of *surface active agent,* a compound that alters surface tension of a liquid in which it is dissolved.

## T

**Terpolymer**—The product of simultaneous polymerization of three different monomers, or of grafting one monomer to the copolymer of two monomers.

**TFE**—Tetrafluoroethylene, a perfluorinated monomer used as a feedstock for the production of PTFE and as a comonomer for the production of a variety of other fluoropolymers.

**Thermoforming**—A process of forming a plastic film or sheet into a three-dimensional shape by clamping it, heating it, and then applying a differential pressure to make the film or sheet conform to the shape of the mold.

***Thermogravimetric analysis* (TGA)**—A widely used method to determine weight change upon heating, such as decomposition, amount of volatile components, including moisture.

**THV**—A terpolymer of TFE, HFP, and VDF.

## U

**Ultraviolet (UV) radiation**—Electromagnetic radiation in the 40–400 nm wavelength region. UV radiation causes polymer degradation and other chemical reactions, including polymerization and cross-linking of monomeric and oligomeric systems.

## V

**Viscoelasticity**—The tendency of polymers to respond to stress as if they were a combination of elastic solids and viscous fluids.

**Viscosity**—The property of resistance of flow exhibited within the body of material. Units of viscosity are Pascal (traditional) and Pa s (SI). Conversion: 10 P = 1 Pa s or 1 cP = 1 mPa s.

**Viscosifying agent**—A substance used to increase the viscosity of a liquid, mainly by swelling.

**Volume resistivity**—The electrical resistance between the opposite sides of a cube.

## W

**Wainscot**—Interior paneling in general and, more specifically, paneling that covers only the lower portion of an interior wall or partition. It has a decorative or protective function.

**Wetting**—The spreading out (and sometimes absorption) of a fluid onto (or into) a surface. In adhesive bonding the wetting occurs when the surface tension of the liquid adhesive is lower than the critical surface tension of the substrate. Good surface wetting is essential for high strength adhesive bond. Wetting can be increased by preparation of the part surface prior to adhesive bonding.

## Y

**Yield deformation**—The strain at which the elastic behavior begins, while the plastic is being strained. Deformation beyond the yield deformation is not reversible.

# Appendix 5: Tedlar Film Designation Guide

The identification codes consist of a combination of uppercase letters and number, such as *TTR10SG3*

- *T* stands for *Tedlar symbol*
- *TR* stands for *end use property or color*
- *10* refers to the *nominal thickness*[1]
- *S* refers to *surface*
- *G* is for *gloss or other properties*
- *3* indicates the *type of the PVF film*

Details regarding the abovementioned letters and numbers are given in the following table:

| End Use Property and Color | Surface | Type | Gloss and/or Other Properties |
|---|---|---|---|
| *FM*—flame modified | *S*—strippable, nonadherable release films | 1—High shrinkage compatible with curing polyester resins | *G*—glossy |
| *MR*—high temperature release | *A*—one side adherable for use with adhesives | 2—High tensile, excellent fold, and flex endurance | *M*—medium gloss |
| *PC*—epoxy board release | *B*—both sides adherable | 3—Standard tensile and elongation | *L*—low gloss |
| *ST*—special transparent | | 4—High elongation, good formability, good heat sealing properties | *S*—satin gloss |
| *TR*—transparent | | 5—Excellent formability—for release and lamination to engineering plastics and metals | *E*—meets aircraft specifications |
| *UT*—ultraviolet screening film, transparent | | | |
| *UW*—ultraviolet screening film, translucent white | | | |
| *WH*—white | | | |

*Source*: TEDLAR Polyvinyl Fluoride Film, Film Designation Guide, H-49723, DuPont, www.dupont.com, 2019.

[1]Nominal thickness or approximate gauge thickness times 10; for example, 10: nominal 100 gauge = nominal 1 mil.

# Bibliography


B. Ameduri, B. Boutevin, Well-Designed Fluoropolymer Synthesis, Properties and Applications, Elsevier B.V, Amsterdam, 2004.

J.G. Drobny, Fluoroelastomers handbook, Definitive User's Guide and Databook, second ed., Elsevier, Oxford, 2016.

J.G. Drobny, Fluoroplastics, Report 184, vol. 16, No. 4, Rapra Technology, Shawbury, Shrewbury, Shropshire, 2006.

J.G. Drobny, Specialty Thermoplastics, Landolt-Börnstein, Group VIII, vol. 13, Springer-Verlag, Heidelberg, 2015.

J.G. Drobny, Technology of Fluoropolymers, second ed., CRC Press, Boca Raton, FL, 2008.

S. Ebnesajjad (Ed.), Introduction to Fluoropolymers: Materials, Technology, and Applications, Elsevier, Oxford, 2013.

S. Ebnesajjad, Expanded PTFE Application Handbook, Technology, Manufacturing and Applications, Elsevier, Oxford, 2017.

S. Ebnesajjad, Fluoroplastics, second ed., Non-Melt Processible Fluoropolymers, The Definitive User's Guide and Databook, vol. 1, Elsevier, Oxford, 2014.

S. Ebnesajjad, Fluoroplastics, second ed., Melt Processible Fluoropolymers, The Definitive User's Guide and Databook, vol. 2, Elsevier, Oxford, 2015.

S. Ebnesajjad, P.R. Khaladkar, Fluoropolymers Applications in Chemical Processing Industries, The Definitive User's Guide and Databook, William Andrew Publishing, Norwich, NY, 2005.

S. Ebnesajjad, Polyvinyl Fluoride – Technology and Applications of PVF, The Definitive User's Guide and Databook, Elsevier, Oxford, 2013.

S. Ebnesajjad, Surface Treatment of Materials for Adhesion Bonding, William Andrew Publishing, Norwich, NY, 2006.

G. Hougham, P.E. Cassidy, K. Johns, T. Davidson (Eds.), Fluoropolymers 1, Synthesis, Kluver Academic/Plenum Publishers, New York, 1999.

G. Hougham, P.E. Cassidy, K. Johns, T. Davidson (Eds.), Fluoropolymers 2, Properties, Kluver Academic/Plenum Publishers, New York, 1999.

L.W. McKeen, Film Properties of Plastics and Elastomers, fourth ed., Elsevier, Oxford, 2017.

B.A. Morris, The Science and Technology of Flexible Packaging, Multilayer Films From Resin and Process to End Use, Elsevier, Oxford, 2017.

J. Scheirs, Fluoropolymers: Technology, Markets and Trends, Rapra Technology Ltd, Shawbury, Shrewsbury, Shropshire, 2001.

J. Scheirs (Ed.), Modern Fluoropolymers: High Performance Polymers for Diverse Applications, John Wiley & Sons, Ltd, Chichester, 1997.

L.J. Wall (Ed.), Fluoropolymers, Wiley-Interscience, New York, 1972.

# Index

*Note*: Page numbers followed by "*f*" and "*t*" refer to figures and tables, respectively.

CPI Antony Rowe
Eastbourne, UK
January 27, 2020